The H

GUIDE

to Hiring

CONTRACTORS

The Homeowners' GUIDE to Hiring CONTRACTORS

How to Save Time, Money and Headaches by Hiring the Right Contractor for Your Job

Brett P. Kennelly
with Eddy Hall

INFORMATION SERVICES GROUP

Cover design by Mirthworks Company, Honolulu, HI
Editorial services by Eddy Hall, Goessel, KS
Page design and layout by Good Shepherd Publications,
Hillsboro, KS

Printed in the United States of America

Library of Congress Catalog Number 97-95347
ISBN 0-9651979-8-0

Disclaimer

The author and publisher are not engaging in and do not intend to engage in rendering legal, financial, managerial, accounting, or other professional services. The material in this book should be considered informational and used as a general guide. Should you require these or other professional services, the assistance of a competent professional should be engaged.

Hiring construction contractors may involve serious and substantial personal and financial consequences. It is not possible nor is it the intention of the publisher or the author to present comprehensive information on this subject. You are encouraged to educate yourself in various ways regarding the hiring of construction contractors, including reading other available materials relating to this subject.

Although efforts have been made to ensure the accuracy of this publication, it is possible that mistakes may exist in both content and typographical form. The information contained in this publication is only current up to the date of printing.

The author, publisher, or agents of the same assume no liability nor responsibility to any person, persons, entities, or organizations with respect to any loss or damage caused, or alleged to be caused, directly or indirectly by the information in this book.

If you do not wish to be bound by the above, you are advised to disregard the information contained in this book and to return the book to the place of purchase.

No index

CONTENTS

Dear Reader,

Almost everyone who has ever completed a construction or remodeling project could tell how he or she would do things differently given the chance. So often people say, "If only I had known before I started what I know now...."

Your level of education, contacts, money, or even negotiating ability has surprisingly little to do with whether your project will be successful. What separates those owners whose construction projects are completed on time and in budget from those who experience nothing short of a nightmare is the specialized knowledge they have before starting their projects.

You can choose to take your chances and go it alone and likely accumulate some stories of frustration and financial loss of your own, or you can invest a few hours and gain the specialized knowledge and tools you need to make your next construction or remodeling project all that you hope for. In the following pages you will receive in a step-by-step format the information you need to make wise choices about one of the most important factors in the success of your construction project. I hope you will make good use of it.

Best wishes for your construction project,

Brett Kennelly

To Susan,
for your perseverance through the desert

INTRODUCTION

WHY DREAMS TURN INTO NIGHTMARES

(and How to Keep It from Happening to You)

You stand on the threshold of one of the biggest investments of your lifetime—building a new custom home or remodeling your present home. As the homeowner, you will be responsible for a wide range of decisions and responsibilities including the process of hiring and working with a construction contractor. The decisions you make will shape the outcome of your construction project and have a significant impact on your personal and financial future.

Though most home construction projects turn out at least reasonably well, frustration and unpleasant surprises are all too common—cost overruns, jobs falling behind schedule, poor quality work, tensions between homeowners and contractors, and sometimes outright fraud. These things can and do happen to virtually anyone, regardless of their level of education, income, or professional background. Why do construction projects often turn into nightmares, or at least disappointments?

More often than not, the customer lacks the specialized

knowledge needed to make the best choices when evaluating and hiring the services of construction professionals. Uncertainty about how to locate and evaluate qualified contractors, how to negotiate a contract that protects their interests, and how to establish and maintain an effective working relationship with a contractor prevents the homeowner from receiving the best services possible.

From *The Complete Guide to Hiring Contractors* you will learn in a few short hours knowledge that most people learn only through years of trial and error, knowledge that will save you time, money, and headaches. Drawing on my career as a building contractor and consultant working with hundreds of homeowners, I'll show you how to take the complex process of hiring and working with a contractor and simplify it by breaking it down into manageable steps and then showing you how to succeed at each step. This easy-to-read guide takes you step by step through the process of locating, screening, and interviewing the contractors who are most qualified for your project. You will learn how to get competitive and thorough estimates and bids from contractors while reducing the chances for cost overruns. You will learn what to look for in construction contracts and how to spot and prevent potential problems before they happen. And you will gain valuable insights about how to develop a win-win relationship with your contractor. By following these steps, you will dramatically improve your chances for a successful construction project.

Even if you have already taken some of these steps—if, for example, you have already found the right contractor—this book can save you money by helping you work more effectively with your contractor and avoid misunderstandings that often cost hundreds or even thousands of dollars. Just as important, the information in this guide can save you valuable time and the grief that can come from making serious but avoidable mistakes.

Each year thousands of homeowners will hire contractors and enjoy the satisfaction of a successful construction project, while many others will be added to the list of those who experience frustration and financial hardship due to a mismanaged project. You don't need to leave the outcome of your project to chance. With the odds of experiencing problems during a construction project so high, and with the cost of problems often running into the thousands and tens of thousands of dollars, the time you dedicate to educating yourself before you start construction will be one of the best investments you ever make. Your return on that investment will likely be a rewarding building experience and, in the end, a beautifully completed construction project of your dreams.

Chapter 1

Knowing What to Expect

The Key to Avoiding Surprises

When you ask people about their experiences with building contractors, you're far more likely to hear horror stories than rave reviews. But contrary to the impression these horror stories might give you, building or remodeling really can go smoothly. It happens all the time. What is the secret of a successful project? It is knowing what can cause problems during a construction project and taking precautions to prevent them.

Throughout this book, stories are used to illustrate specific points. Although names and circumstances have been changed, the stories are based on actual events and in most cases show situations where misunderstandings and therefore problems commonly arise. The following stories provide a few examples of the kinds of problems you might encounter and how you might be able to prevent them.

A Case of a Buried Surprise

Headache: When the Forsters' contractor started digging to lay the footing for the addition to their home, she discovered unexpected soil conditions that required a more expensive construction for the foundation. The Forsters did not feel they should pay the additional expense since they had awarded the contract based on a bid that did not include it. They felt the builder had had every opportunity to examine the soils report, and that they should be able to rely on the builder's bid. The builder claimed that she had no reason to make an allowance for the soil condition discovered because a recent soil test had given no indication of the problem.

Cause: Because the Forsters' contract did not spell out who would pay to correct hidden conditions, neither the contractor nor the homeowners felt responsible to pay the unexpected expense.

Prevention: Hidden conditions are common in the construction business—termite damage, dry rot, underground structures or conditions. A good contract will spell out who will pay to correct any hidden conditions discovered.

In this case, reasonable efforts had been made to determine the soil conditions. Not all jobs require soil tests. In fact, most single family residences do not. When soil tests are performed, test borings or holes are drilled at various points throughout the property, but it is impractical to drill enough holes to eliminate all uncertainty about the soil conditions.

During negotiations between the owner and the contractor, had the possibility of unknown conditions been discussed, the owner could have chosen to pay for more test borings to reduce the chance of surprise. If these additional borings had revealed the unstable conditions, the contractor would then have adjusted her estimate to include the work necessary to overcome the problem. On the other hand, if the owner and contractor had decided not to invest in more soil testing, the contractor would not have inflated her estimate to cover a problem that might or might not exist.

In the story, the dispute arose not because the owner was being unfairly charged, but because the cost came as a surprise to the owner. Avoiding surprises and maintaining trust is critical for an effective working relationships between homeowner and contractor.

A CASE OF CLASHING EXPECTATIONS

Headache: A few weeks into her remodeling project, Merilee's contractor started bugging her for brand and model numbers for light fixtures and bath faucets and asking her what tile and molding patterns she wanted. Merilee had neither the time nor expertise to make these decisions, nor had she given much thought to who should make these decisions or how. Since the contractor had her drawings and plans and Merilee had signed a contract based on those plans, she assumed her contractor should take it from there. Why didn't the contractor buy the light and plumbing fixtures and leave her out of it?

Cause: This problem was caused by clashing expectations. Merilee was frustrated because she expected her contractor to do things that are normally the homeowner's responsibility. Although the contractor had included an allowance for these items in the contract, he had not received the exact specifications for these items. The contractor needed Merilee to make her selections before he could order the materials.

Prevention: In this case the contractor should have reviewed with the owner prior to construction which decisions the owner would need to make and by what date. Although most construction drawings include substantial detail, often the color, finish, and specific model numbers are not included for such items as moldings, light fixtures, plumbing fixtures, hardware, and appliances. Even when they are included, what is specified in the drawings and what is ultimately installed often vary because of discontinued items, the need to reduce costs, or changes requested by owners.

Although most builders are more than willing to offer their opinions on aesthetic decisions, few are willing to make the actual decisions and usually should not, since it is the owners who will have to live with the choices.

If Merilee did not have the time or confidence to make decorating decisions, she could hire an interior decorator to help her. Or perhaps a friend who was good with color and design would be willing to give her some advice.

The time to determine how such choices will be made is before the contract is signed. This gives you time to arrange for any needed additional help and prevents delays and misunderstandings.

It may help to think of your contractor as being more like a cook than a chef. Cooks follow recipes created by others. A chef, on the other hand, not only cooks, but also creates the recipe. Unless you've hired your contractor to participate as a chef by becoming involved in the design and decorating aspects of the project, to assume that your builder will do this may lead to frustration and misunderstanding. In reality, most builders contribute to the design process to some degree, either formally or informally, but it is important to understand the difference between the two roles.

A CASE OF WORK GRINDING TO A HALT

Headache: After getting off to a good start, Eric's kitchen remodel ground to a halt. He complained to his general contractor that no one was working on his kitchen. The builder explained that they were waiting on cabinets. That seemed like a poor excuse to Eric; it had been four weeks since he told the contractor what cabinets he wanted. Besides, why couldn't the crew work on other parts of the job while they were waiting for the cabinets?

Cause: Many materials, such as cabinets, must be ordered several weeks or even months in advance. When they are not ordered on time, not only does this delay installation of the cabinets, but it also holds up installation of the flooring, countertop, plumbing fixtures, and appliances.

Prevention: To keep this from happening to you, allow plenty of lead time in making your selections of materials and products. To avoid delay, the contractor should have made it clear to Eric that the cabinets needed to be ordered six to eight weeks before they were actually needed on the job. Your contractor can advise you on how much lead time is needed for each decision you will be making.

A Case of Final Price Shock

Headache: Ten or twelve times during the building of their addition, the Friesens asked their builder to make changes in their original plans. They knew the changes would involve some expense, but when at the end of the project they received a bill for $13,000 to cover the changes, they were shocked and felt the builder was taking advantage of them. They balked at paying the bill.

Cause: The bidding process rewards the bidder with the lowest bid, creating an incentive to bid jobs artificially low—often at or sometimes even below the contractor's cost. As a result, some—in fact, many—builders lowball their bids with the hope they can make their profits from change orders. They fear, many times justifiably, that if their original bid reflects the actual cost of completing the project, they will not get the job.

Even when the contractor is scrupulously honest, a homeowner should not blindly request changes, assuming they will be affordable. It doesn't take many seemingly small changes to add up to a substantial sum.

Prevention: You can do several things to keep this from happening to you.

First, do not authorize any changes until the contractor gives you a change order to sign that specifies exactly how much the change will cost. That way if you find out it will cost $336 to have a window moved and it's not worth that much to you, you can decide to leave the window where you had it. Once the work has been done, it's too late to decide it wasn't worth the price.

Second, have your contractor spell out in the contract exactly how charges for changes will be calculated. What will be the hourly charge for journeyman labor? For mid-level skill labor? For unskilled labor? What will the contractor's markup be?

Third, occasionally you will want or need a change for which the contractor will not be able to give you an exact estimate. This might occur, for example, with termite or water damage, or with work that is unique or nonstandard. In such situations, have your contractor give you a written approximate cost to perform the work with a not-to-exceed number. Do not proceed with the work unless you receive reasonable assurance from the contractor that the work being considered will not be open-ended once started. These precautions minimize your chances of getting a surprise bill for extra work.

Fourth, verify that the specifications for your project are detailed. What kinds of doorknobs, faucets, light fixtures, and trim do you want? The more specific you are, the less room there is for the price to change once construction begins. Is this a lot of work? Yes, it is. Does it pay off? You bet it does.

If you already know which contractor you're going to use, one way to avoid inflated charges for changes is a *negotiated contract.* Rather than the usual approach of your getting competitive bids from general contractors, the contractor agrees to a predetermined fixed fee for his services and gets competitive bids from subcontractors and suppliers. Under this arrangement, your builder normally will not receive any additional profit from change orders. This is discussed in greater detail in the chapter on contracts.

These are just four of dozens of hassles you might encounter during your building or remodeling project. Common occasions for misunderstanding between owners and contractors include:

✓ Who will make certain decisions and when they are due

✓ Regularity of job site clean up

✓ Responsibility to pay for overcoming or correcting unanticipated problems

✓ Days and times the contractor will be working on the job

✓ Number of carpenters the contractor will dedicate to the job to complete the work

✓ Who is responsible for moving the owners' personal property

✓ What areas or facilities, such as bathrooms and kitchens, the owners will be able to use during construction

- ✓ Amount of time the contractor will personally be on the job site
- ✓ How easily and during what hours you can reach your contractor
- ✓ When payments will be made and in what amount
- ✓ Liens and lien releases
- ✓ Quality of craftsmanship

Having realistic and well-defined expectations of your contractor before the project starts will prevent a lot of headaches for both you and your builder.

This book is, in effect, a headache prevention manual. Since a positive building experience depends primarily on minimizing problems, as you go through this book you will learn about many more potential headaches and how to avoid them. Perhaps the most common, and potentially the most serious, mistake of all is choosing the wrong contractor for your job. The next few chapters show why this mistake is so common and outline steps that can lead you to the best contractor for your project.

CHAPTER 1
AT A GLANCE

✎ Misunderstandings about which responsibilities are the contractor's and which are the homeowner's are common.

✎ A contractor is more like a cook than a chef. Unless you specifically hire a contractor to help design your project, assume that the contractor will be working with the instructions and ingredients provided by others—not making design decisions.

✎ Know what services your contractor will not be providing. Arrange to provide these yourself or to hire professionals to do them.

✎ The key to a successful construction project is to anticipate potential problems and take precautions to keep them from happening.

Chapter 2

How to Find Potential Contractors

When it came time to hire a contractor for a major remodel on his home, Stewart carefully followed all the advice he had heard: Ask people to recommend contractors they have been happy with, get three bids, and avoid the high and low bids.

Stewart's architect had recommended a builder. That was one. An associate at work had recently finished a remodeling project and spoke highly of his contractor; that made two. And Stewart had recently had his offices remodeled and liked the contractor's work; that made three. Stewart felt confident he had the names of three good builders.

He asked all three to bid his project. Since the job was big, he expected the bids to be hefty. He didn't, though, expect them to be so far apart. They ranged from $275,000 to $490,000. Because he had heard that the low bid can sometimes mean trouble, Stewart accepted the middle bid of $375,000, the one submitted by the builder who remodeled his offices.

The quality of the builder's work turned out to be okay, but he didn't get the work done on schedule, and it came in far over budget.

Stewart had been so careful to do everything right. How, he wondered, could things have gone so wrong?

THE SINGLE GREATEST KEY TO SUCCESSFUL BUILDING

One decision determines the success or failure of your construction project more than any other: hiring the right contractor. Just as a symphony's conductor is the key to the orchestra's performance, the experience, personality and managerial skills of the general contractor, more than all other factors, will determine the success or failure of your construction project.

In spite of this, most people spend more time shopping for a VCR than they do shopping for a contractor. Why? Because no one has ever told them how to shop for contractors. Yet knowing how to shop for a contractor is the one skill most critical to the success of your construction project. This is true because no matter how detailed your drawings and specifications, no two contractors will do your job the same way.

Choosing a contractor is like choosing a restaurant. If you go to three Italian restaurants and order exactly the same entree at each—spaghetti and meatballs—the quality, price, and service will be different at each one. Since a construction project involves many more variables than a simple meal, the quality of work, the price you pay, and your level of satisfaction with the service can vary widely depending on which contractor you hire. Just as you can decide which Italian restaurant you most enjoy (though other diners may prefer a different one), you can learn to identify the builder best qualified to do *your particular job*—one who will give you top-quality work and excellent service at a fair price.

DOES THE "GET THREE BIDS" RULE WORK?

Stewart followed the conventional wisdom of getting three bids and hiring the contractor who turned in the most attractive bid. Sometimes this works out, but more often, as Stewart found out, it does not.

Does this mean "get three bids" is bad advice? No, it's sound advice, but advice often misunderstood and misapplied. It was not a mistake for Stewart to get three bids; his mistake was in asking for bids too soon—before he had a list of contractors specifically qualified to do his particular job.

In Stewart's case, only one of the three contractors was right for the job—the one recommended by the architect. That firm had done many projects similar to Stewart's and knew what the job would take. The low bid came from a builder who had never done a job larger than $75,000. He simply didn't have the experience to prepare an accurate bid for a job of that size. The contractor Stewart chose, the one who had done a great job remodeling his office, had limited experience with residential remodeling. That's why he so badly underestimated how long the project would take.

Stewart's contractor list was much too short. Because he didn't realize that, and because it didn't include enough qualified builders, he hired the wrong person.

Before you ask builders for estimates, first make a list of ten to fifteen contractors. Then check out each of those contractors, determining which are qualified to do your work. Many won't be. From the remaining short list, you will then ask three well-qualified builders to work up estimates for your project.

COMMON WAYS TO FIND CONTRACTORS

Hundreds of builders may be willing to bid on your project, but most of them won't be right for it. Some ways of finding contractors are more likely than others to lead you to builders well-matched to your needs.

Advertisements. An obvious way to find contractors is through advertising. Look in the phone book. Check your local paper.

The problem with this approach is that many builders don't advertise. If you depend primarily on advertising to create your list, you will overlook some of the best builders in your area.

Friends and relatives. Most people have friends, relatives, or acquaintances in the construction business. But simply because a contractor is a friend, attends the same church, or seems like a nice person does not mean the builder is qualified to do your project.

Jeanie's brother-in-law, Bill, was a contractor, so when Jeanie needed to reroof her house, she was glad to be spared the headache of looking for someone to do it. Bill put on the cedar shake roof for her and it looked great.

Four months later, the rainy season started and Jeanie noticed a leak in the hall closet. She called Bill and he came over and fixed it. Or thought he did.

The leak kept coming back. So Jeanie got a second opinion. The second contractor, a roofing specialist, told Jeanie that parts of her roof flashing had been installed incorrectly and needed to be redone. It cost her $600 to correct the problem. Since Jeanie hated confrontation, she was reluctant to tell Bill what had happened. As a result, at the next family get-together, Jeanie was uncomfortable around Bill.

Hiring a friend or relative to do your construction can lead to unwelcome complications.

Form 2.1

Contractor List

Questions to ask when gathering names of contractors:

Do you know of any contractors who specialize in ______________ (your type of project)?

Have you worked with this builder? How many times?

Were you satisfied with the quality of the contractor's work and service?

Would you hire this builder again?

Do you know of anyone who has recently done a similar project?

List any contractors you want to check out.

	Company or Name	License #	Phone #	Referred by
1.				
2.				
3.				
18.				
19.				
20.				

- ✓ Many people hire family or friends hoping to get a "good deal." Builders who give "good deals" tend to lose interest in projects when they are not making money.

- ✓ If you hire an attorney to hold a friend or relative accountable, you may jeopardize personal or family relationships.

If Jeanie had hired someone outside her family, she might have been more persistent with the contractor about correcting the problem. In fact, if, rather than hiring Bill because he was family, she had considered Bill's qualifications along with those of other candidates, she would have discovered that Bill was not a roofing specialist and therefore not a good choice for her job. The truth was, because Bill didn't specialize in roofs, he really didn't want the job. But he felt obligated to take it because Jeanie was family. Some good, honest communication could have saved both Jeanie and Bill some grief.

If you believe not hiring a certain person might create hard feelings, meet with the person and explain your concern. Explain that your relationship is important to you and that you want to avoid any circumstances that could jeopardize it.

Hiring a builder you have worked with in the past and with whom you have since become friends is different. Each party enters the relationship knowing what is expected. The homeowners come into the situation with their eyes open. This situation usually produces a highly effective working relationship.

If you still want to consider hiring a friend or relative, the secret to making this work is to apply the same hiring principles to family or friends that you apply to strangers. Only hire the relative or friend if that person would win the bid even without the personal relationship.

BEFORE YOU START YOUR LIST

So, if listing contractors you know or whose ads you have seen is not good enough, how do you go about it? A good way to find contractors is to start by listing people who may know building contractors. Your list can include neighbors, relatives, coworkers, and friends.*

People who have had work done by contractors and who are very happy with the work done are usually the best sources. They know, for example, if the builder started and completed the work on time, performed quality work, and resolved problems fairly and efficiently.

People who work with contractors can also be an excellent source for names of good builders. The work of drywall contractors and cabinetmakers who install cabinets, for example, is directly affected by the quality of contractors' work. Both the drywall contractor and the cabinetmaker can tell you in a minute who the better builders are. The framing in a home built by a quality builder is more accurate, making the subcontractor's work easier. Better contractors also pay their bills on time, and subcontractors prefer to work with contractors who do quality work and pay their bills.

Any of the following may be able to refer you to good builders:

- ✓ Architects
- ✓ Bankers
- ✓ Building designers

---

* *See Appendix A for a reproducible master of this form.*

- ✓ Interior decorators
- ✓ Cabinet shops
- ✓ Local chapters of trade organizations
- ✓ Lumberyards that specialize in working with contractors
- ✓ Real estate agents
- ✓ Subcontractors

CREATING YOUR LIST

Once you have a list of people who may know contractors, contact them and ask if they can recommend any general contractors who are experienced with the type of work you are planning. Use the questions at the beginning of Form 2.1, "Contractor List," as a guide. A reproducible master of this form and all other forms mentioned in this book is found in Appendix A. Most people will be more than willing to give you their opinions.

As people make recommendations, record the names of the builders along with their phone numbers, who referred them, and any other information you get. If some people on your list don't know any general contractors, ask if they know others who might. Add those names to your first list and contact them as well.

If you find yourself wondering if it is worth all this effort to generate a long list of contractors, ask yourself how much you want to avoid playing the starring role in a construction horror story. Remember: Many contractors do little or no advertising. Unless you take the initiative to actively look for builders, you may never find the best one for your job.

How Long a List?

How do you know when your list is long enough? Once you have the names of ten or so contractors others have recommended, I suggest you start screening them in the way described in the next chapter. If out of those ten you come up with three or four strong candidates, you may not need any more names. But, if only one or two builders impress you enough that you want to interview them, go back and collect more names. Keep gathering names and checking them out until you have found three or four contractors that have been highly recommended and are experienced with the type of work you are considering.

You may end up with a "long list" of ten to twenty names, but if that's how many you need, it's worth it. If the best contractor for your job isn't on your long list, he or she can't show up on your short list. So keep looking until you have found at least three or four strong candidates. Only then is your list long enough.

Chapter 2
At A Glance

✎ One decision determines the success or failure of your construction project more than any other: hiring the right contractor.

✎ The quality of work, the price you pay, and your level of satisfaction with the service you receive can vary widely depending on which contractor you hire.

✎ Many of the better builders can be hard to find because they often do not need to advertise.

✎ Use caution when hiring friends or relatives.

✎ To find a good builder, broaden your search by including the eyes, ears, and experiences of others.

✎ The two best sources for names of contractors are people who have had construction work done and are very happy with the work, and people who work with contractors—such as architects and real estate agents—and whose jobs are impacted by the quality of the contractors' work.

✎ You may need to check out ten to twenty contractors or even more before finding three who are qualified to bid on your project.

CHAPTER 3

PRESCREENING

The Essential First Step

The Howards needed a contractor to do a major remodel job on their home. They knew of a builder who was working on another home in their neighborhood, so they went to see his work. Though the job had just gotten started, the demolition work was coming along well and the owners were happy with the work.

So the Howards met with the contractor one evening over coffee. He showed up on time, professionally dressed. He gave them an attractive brochure and seemed to know his stuff. The Howards were impressed and asked him for a bid.

The bid turned out to be good yet not unrealistically low. They accepted the bid and made a substantial down payment.

Soon the work started, and the first few weeks were exciting—men working, trucks coming and going. They were finally on their way! After a couple of more weeks, though, things slowed down. The demolition was finished, but the rebuilding wasn't starting.

When the Howards called the contractor, he assured them the work would start up again soon; the crew was just waiting on materials. Out of curiosity, the Howards checked with the other family in the neighborhood using the same contractor. The contractor had been telling them the same story, only for longer. When the Howards tried to call the builder again, they got a recording saying the number was not in service. Further investigation revealed that the builder had skipped town leaving behind several other unfinished projects and taking substantial sums of customers' money.

CONTRACTORS TO AVOID

Sound impossible? It happens. It happens to all kinds of people, no matter how professional or well-educated they are. The Howards could have prevented this nightmare if they had prequalified contractors before deciding which ones to interview. Once you have your long list of contractors, your next step is to prescreen all the companies on your list. With a quick but careful prescreening, you can do a pretty thorough job of cutting the following three kinds of contractors from your list.

- ✓ *Con artists.* As the Howards learned too late, some people posing as contractors are actually professional con artists. They move from town to town starting business after business, using new names and license numbers each time. A simple background check is often all it takes to tip you off to such a scam.

- ✓ *Unqualified contractors.* All builders do not do equally professional work. Some frequently fall behind schedule, go over budget, and do poor quality work. If a builder has earned a reputation for being unprofessional, you can find that out through prescreening.

- ✓ *Mismatched contractors.* Whether they build fences or factories, most firms specialize in a specific type of work. When looking for a contractor, focus on builders that have proven track records with the type of project you are planning. A builder that specializes in kitchen and bath remodeling may not be a good choice for your new custom home. And unless work is scarce, most custom home builders will not be interested in remodeling your bathroom.

Look at the type and price of the majority of the builder's previous projects. Think twice before hiring a contractor for a $250,000 remodeling project if most of the contractor's previous experience is with projects of under $50,000.

Defining Your Minimum Requirements

Your first goal, in Step 3, is to eliminate from your list all those contractors that fall short of your minimum requirements. What should your minimum requirements for a contractor include? A qualified contractor:

- ✓ Has a proven track record with your type of construction in your community
- ✓ Has many highly satisfied previous customers
- ✓ Has a stable financial history
- ✓ Has been in business at least five years
- ✓ Communicates with you openly and effectively
- ✓ Listens and offers suggestions or solutions in response to your construction problems

✓ Is someone you feel comfortable working with

✓ Provides value, not merely the lowest price

Through a two-part prescreening process, you can quickly identify most of the contractors on your list that don't meet these minimum requirements and cross them off your list.

Part 1: The Background Check

Several state, local, and private agencies monitor the licensing of construction companies and complaints filed against them—the Better Business Bureau, your state consumer agency, state licensing boards, your local building department, and any trade associations to which contractors belong. Contact these agencies and ask the questions in Part 1 of the Contractor Telephone Interview Form (Form 3.1) to find out if they have any information about the contractors on your list.

While this precaution can alert you to some unscrupulous or incompetent builders, be aware that it has its limits. Most homeowners who have problems with builders don't report those problems to the monitoring agencies. Plus, it takes time for complaints to be filed, so information on recent complaints may not yet be available. This means that a clean record with monitoring agencies doesn't guarantee that a builder is reliable.

At the same time, just because someone has filed a complaint against a builder doesn't mean the builder is untrustworthy. Misunderstandings occur. Some customers are hard to please and quick to file complaints. More important than the existence of a complaint is what the builder did or didn't do to resolve it. The monitoring agency should be able to tell you how the builder responded to each complaint filed.

PART 2: THE PRELIMINARY PHONE INTERVIEW

Once you've eliminated any firms that don't pass your background check, you're ready to make a quick phone call to each contractor left on your list. With a few questions, you should be able to determine:

- ✓ Whether the contractor specializes in the kind of work you need done
- ✓ How long the company has been in business
- ✓ Whether the firm has proper licensing, insurance, and bonding
- ✓ Whether the contractor comes across as professional and cooperative

Use the questions in the Contractor Telephone Interview form to conduct background checks and preliminary phone interviews.* (See box.) Record the contractors' answers.

When you first call builders, before you mention what kind of project you want done, it can be helpful to find out what projects they have completed in the last twelve months.

Why? When asked if they do a certain kind of work, most builders will say yes, even though they may have only done one such job twelve years ago—or perhaps even if they have never done that kind of work but think they could.

When the builder doesn't know what skills you are looking for, and you ask for a list of recently completed projects, you

---

* *See Appendix A for a reproducible master of this form.*

Form 3.1

Contractor Telephone Interview

Company name ______________________ Phone ______________

Person ______________________________ Date ______________

Address ___________________________ License # __________

____

Recommended by ____________________

STEP 1: BACKGROUND CHECK

Contact these agencies and ask the following questions.

Better Business Bureau

- Number of years in business __________
- Have any complaints been filed? ( ) yes ( ) no
- How many? When? ____________________________
- Were the complaints resolved? ( ) yes ( ) no

Your state agency for handling consumer complaints *(such as attorney general)*

- Have any complaints been filed? ( ) yes ( ) no
- How many? When? ____________________________
- Were the complaints resolved? ( ) yes ( ) no

Local building department

- Has this builder completed any projects in town? ( ) yes ( ) no
- Do you work with this builder often? ( ) yes ( ) no
- Have you had any complaints against this builder? ( ) yes ( ) no

State licensing board for contractors *(if your state requires contractors to be licensed)*

- Is this builder's license current? ( ) yes ( ) no
- Is the license issued to the person or company I am considering hiring? ( ) yes ( ) no
- Does the contractor have a current license bond in good standing? ( ) yes ( ) no
- Have there been any complaints against this license? ( ) yes ( ) no

STEP 2: TELEPHONE PRESCREENING

Conduct a preliminary phone interview with the contractor.

It helps to qualify yourself if you tell the contractor who gave you his or her name. Then ask:

1. What type of construction do you specialize in? (If possible, let the contractor describe what kind of work he or she specializes in before describing your project.) ______________________

   ____

   ____

2. What kind of projects are you working on now? ___________

   ____

   ____

3. What distinguishes your company from other builders? ______

   ____

4. How long have you been in business? __________________
5. Is your office nearby? ______________________________
6. Do you have any completed projects in the area? __________

   ____

7. When can you start new projects? ____________________
8. What do you think I should look for when hiring a contractor?

   ____

9. Do you carry workers' compensation insurance? ( ) yes ( ) no

   Liability insurance? ( ) yes ( ) no

Following the phone interview:

Do you want to schedule a personal interview with this contractor?
( ) yes ( ) no

If so, be sure to schedule the meeting so everyone who will be involved in selecting the contractor can attend.

Date ____________________________

Time ____________________________

Place ____________________________

get a more realistic picture of what the builder truly specializes in. The ideal contractor will specialize in your particular type of work, but other contractors may do your type of work often enough to be qualified for the job.

If a builder becomes impatient or uncooperative with your questions, consider it a yellow flag. It suggests how the builder is likely to respond during the challenges that are sure to arise during construction. Builders who provide quality products and services respect customers who do their homework.

Contractors can be hard to reach by phone. Since most builders are on the job during the day, they are usually easiest to reach in the early morning or evening. If you leave a message, be sure to mention who referred you to the builder and ask the builder to contact you by a specific date.

ANSWERING BUILDERS' QUESTIONS

When you call builders, expect them to ask you questions as well. Why? They want to know if you are the kind of customer they want to work with. Just as you are looking for the best contractor for your job, good builders are always looking for the best projects and customers.

Good builders want to know as much about you and your project as you want to know about them. They don't want to waste time following leads that are unlikely to produce work. Here are some of the questions you should be prepared to answer:

✓ How did you hear about our firm?

✓ When will the project start?

✓ Who will be doing the building designs?

- ✓ What kind of construction is involved in the project?
- ✓ Is the construction financing in place?
- ✓ Are the specifications complete?
- ✓ What is the status of your building permit?
- ✓ How many contractors will be bidding on the job?
- ✓ Who are the other builders? (You may not know yet, but the builder may ask.)
- ✓ What is the project budget?

Customers Builders Avoid

Just as you want to avoid certain kinds of builders, builders avoid certain kinds of customers.

- ✗ *Legalistic customers.* Contractors try to avoid customers who deliberately take advantage of them by looking for excuses to not pay the bill.
- ✗ *Nitpicky customers.* Even if a customer is well-meaning, if he or she comes across as extremely perfectionistic, a builder might be reluctant to take the job. Perfectionistic customers can be hard to please and slow to pay. This bias that most contractors hold can present a challenge to homeowners who want to be appropriately thorough in evaluating potential builders. Your builder needs to know that while you are thorough, you are also reasonable.

✗ *Window-shoppers.* Contractors regularly get calls from people who hope to build someday, but who aren't really serious about it, so builders learn to separate qualified buyers from window-shoppers.

✗ *Price-only shoppers.* Honest builders almost always seem more expensive at first because they submit honest, accurate estimates. Less qualified builders often submit lower bids either because they intentionally lowball to get the job, or because they don't know what they are doing. If a customer is looking for the lowest price in town, a quality builder will not want to invest the considerable time it takes to work up a bid. To interest quality builders, make clear that you are looking for a quality builder, not the cheapest builder you can find.

Between your background checks and your preliminary telephone interviews, you have probably crossed many of the names off your list because they don't meet your minimum requirements. Once you have identified three or four contractors who you believe are well-qualified to do your work, you are ready for Step 4: interviewing your top candidates.

CHAPTER 3
AT A GLANCE

✎ Many of the contractors on your long list will not be right for your job.

✎ You need to screen out three kinds of builders: con artists, unqualified contractors, and mismatched contractors.

✎ The first step in prequalifying contractors is to conduct background checks by calling agencies that monitor construction companies.

✎ The second step in prequalifying contractors is to conduct a preliminary phone interview with each contractor.

✎ When you call builders, expect to answer their questions as well.

✎ Once you have identified three or four contractors who meet your minimum requirements, you are ready to move onto the next step—the personal interview.

CHAPTER 4

HOW TO INTERVIEW YOUR TOP CANDIDATES

When Jackson and Shirley Landrum, a professional couple, hired their contractor, they had conscientiously done their homework. They had asked friends, coworkers, and people in their local community for the names of good builders. They had conducted background checks and preliminary phone interviews and whittled their long list of contractors down to a short list. A very short list, in fact.

One builder stood out far above the rest. Not only was he well-respected in the community, but he specialized in the just kind of work they needed done. By the time the Landrums met with him for the formal interview, they were no longer wondering if he was the builder for them; they just hoped he would take the job.

To their delight, he did take the job. But a year later the Landrums were suing their dream contractor as they prepared to have much of the work that had been completed redone.

What went wrong? Because the Landrums were convinced, even before they interviewed him, that this was the right contractor for their job, they failed to ask certain critical questions during the interview. The right questions would have revealed that this contractor was already overcommitted and that their job would be supervised by a newly hired and underqualified project supervisor.

The combination of the supervisor's lack of experience and the contractor's inability to properly oversee the job resulted in much of the work being done incorrectly. A few pertinent questions during the interview could have saved the Landrums two years' worth of headaches and tens of thousands of dollars.

During the interview, the contractor had impressed Roland with his extensive knowledge of construction and his photo portfolio of completed projects. Roland had hired the builder, confident of his high qualifications for the job.

But now that the builder was on the job, Roland dreaded showing up at the job site. Whenever Roland asked him a question, the builder broke out into a string of profanities. "You're wasting my time with all your stupid questions," the builder would say. "You're making me lose money." When Roland kept asking questions and requesting information, the contractor just got hotter and threatened to walk off the job.

Feeling the contractor had him over a barrel, Roland eventually gave in to the contractor's tactics, hoping to just get the job done rather than having to deal with all that would be involved in firing the contractor and finding someone to replace him. The contractor did finish the work, but the completed job did not include many of the features Roland had wanted.

Could any of this have been avoided? Looking back, Roland realized that when he had met with the contractor, he had focused almost entirely on just getting a bid. He hadn't taken time to get to know the contractor to find out how easy or difficult it would be to work with him.

The Purpose of the Interview

Your first personal interview with a contractor has a simple yet critical purpose—to learn as much as you can about the builder's qualifications and personality. As these stories show, an interview that is mishandled or skipped over lightly can spell disaster for your construction project. A well-handled interview can lead to a strong, cooperative working relationship between you and your contractor.

In most cases, this interview will be your first chance to see how the contractor conducts business. Does the builder show up on time? Does he or she listen and answer your questions clearly? Does the builder seem glad to explain things or too busy to be bothered with your concerns?

If you will be working with the contractor personally, your ability to work together is crucial. Staying on friendly terms during the early stages of a project is fairly easy. Keeping communication open during disagreements can be a challenge. The personal interview gives you a chance to gauge whether you and the contractor are likely to clash or to work well together.

Making Appointments

When scheduling appointments, keep the following guidelines in mind.

Where? If possible, hold the meeting at the location of the proposed construction.

Who? Everyone who will be involved in selecting the builder should attend.

How long? Allow 1fi hours or more for the meeting.

What? Ask the builder to bring:

1. Evidence of current workers' compensation and liability insurance

2. Proof of contractor's license

3. Photos of previous work (if available)

4. Eight to ten recent customer references

Preparing for the Meeting

A few simple preparations can pave the way for an effective interview.

1. Make arrangements to prevent potential distractions—phone calls, pagers, children, pets, the TV, your secretary, or an unfinished evening meal.

2. Have your list of questions ready. If you plan to use the Contractor Personal Interview form, have it ready and fill in any information you already know. (See box for possible questions. A reproducible master of the Contractor Personal Interview form is in Appendix A.)

3. Highlight any questions especially important to you.

4. Bring preliminary or final drawings, if available, for the contractor to review. Pictures from magazines or brochures of products and architectural styles you like can also be helpful.

Questions From Form 4.1

CONTRACTOR PERSONAL INTERVIEW

PRELIMINARY QUESTIONS

1. Can you tell me about your background and how you got into construction?
2. How long have you worked in the construction field?
3. What do you like the most and least about being a general contractor?
4. How much time do you spend on the job and what is your role in the construction?
5. What distinguishes your company from other contractors that provide similar service?

PRIMARY QUESTIONS

1. Do you like to become involved in the construction design process? Why?
2. What are the keys to a successful construction project?
3. What are some common problems contractors face when working with owners?
4. What are some common problems owners face when dealing with contractors?
5. How can we avoid these problems?
6. How do you ensure that the project is completed on time?

Questions From Form 4.1—continued

7. What can we do to keep the final bids within our construction budget?
8. Once we start the project, how do you ensure that our project stays within budget?
9. How do you price requests for changes or additional work?
10. How do you handle complaints or misunderstandings during the construction project?
11. Do you subcontract any of the work? Why?
12. Who supervises the work?
13. How long have you worked with the subcontractors you use?
14. How do you keep your contractors accountable to the schedule and quality standards?
15. How do you ensure the quality of your work?
16. How is your payment policy structured?
17. How can we be sure the general contractor is paying the subcontractors?
18. Under what circumstances should we withhold payment?

CLOSING QUESTIONS

1. Do you have any photos of previous projects we can see?
2. Do you have a list of references you can leave with us?
3. Can we see some of the projects you have completed?

Questions From Form 4.1—continued

4. Do you have any projects currently under construction that we could visit?
5. Do you have a sample of your contract with you?
6. Do you provide preliminary estimates?
7. How available are you when we need to contact you during the project?
8. How much variance is there between your preliminary estimates and your final bids?
9. What advice can you give us on how to choose a contractor?

CONDUCTING THE INTERVIEW

The keys to learning as much as possible about the contractor during the interview are asking the right questions, listening carefully, and observing how the contractor reacts. Because construction projects involve coordinating the efforts of many people, a good general contractor must be able to communicate effectively and give attention to detail.

Contractors' answers to your questions will either build confidence or raise doubts about their technical abilities and business management skills. While there are no perfect contractors, any good contractor should be able to give answers that are relevant and easy to understand. When contractors don't know the answers to questions, note how they handle it. Do they gloss over the questions? Change the subject? Or do they say, "I don't know, but I'll try to find out"?

During the interview, keep the following pointers in mind:

✓ Remember: The purpose of the interview is to learn as much as you can about the builder's skills, reliability, and personality.

✓ Just because someone is willing to give you a bid doesn't mean you should ask for one.

✓ Try to get answers to your questions. Let the contractor do most of the talking, but don't let him or her talk too long about one topic, especially once your question has been answered.

✓ Don't be intimidated. Avoid letting the builder control the meeting.

✓ Notice whether the builder seems more interested in selling you a job or in helping you with your construction needs.

Recording Your Impressions

The best time to record your thoughts and impressions about a builder is right after the interview. Use the Contractor Evaluation form to record the strengths and weaknesses of each candidate. Answer Questions 11 through 13 only after talking with the contractor's previous customers.*

Checking References

Although many people consider references provided by a builder to be inherently biased and therefore not very helpful,

* *See Appendix A for a reproducible master of this form.*

FORM 4.2

CONTRACTOR EVALUATION

Use this tool to evaluate each contractor you interview in person. After the personal interview fill out Part I. After calling the contractor's references fill out Part II. Finally, fill out Part III. Circle the number that goes with the answer that is most nearly true for each question.

Contractor's name ______________________________

Date of personal interview ______________________________

Names of references checked ______________________________

____

____

____

Part I: Interview

THIS CONTRACTOR:	FALSE	UNSURE	TRUE
1. Arrived on time for the appointment.	0	1	3
2. Was well-prepared for the meeting.	0	1	3
3. Listened and responded to what I said.	0	1	3
4. Gave satisfactory answers to my questions.	0	1	3
5. Seemed to understand my ideas.	0	1	3
6. Acted interested in my project.	0	1	3
7. Has completed similar projects in the area.	0	1	3

8. Brought the information I requested.	0	1	3	
9. Never became impatient or irritated.	0	1	3	
10. Avoided criticizing previous customers, employees, and other firms.	0	1	3	

Part II: References

Talk with at least six of the contractor's references before answering this question.

	No	*Somewhat*	*Yes*	*Very*
11. Are this contractor's previous customers satisfied with the quality of work and service they received?	0	10	20	30

Part III: Gut Feelings

	No	*Somewhat*	*Yes*	*Very*
12. Do you think you would be comfortable working with this person on a daily basis?	0	5	15	20
13. Do you have a favorable overall impression of this builder?	0	5	15	20

TOTAL OF CIRCLED NUMBERS (100 possible points): [______]

Note: This tool is one guide to help you evaluate contractors. Factors not listed here may be relevant to your decision, and you may wish to give some factors more or less weight than they are given here.

the right questions can yield a wealth of useful information. The keys to getting the information you need are: (1) having enough references and (2) knowing what questions to ask.

Three customer references are simply not enough. Even incompetent builders can find three people who will say good things about them. The contractor you hire should be able to give you the names of eight to ten recent customers. If not, find out why not. Has the contractor just moved to the area? Is he or she a new builder? Or does the company not have that many satisfied customers?

Call at least six references, preferably more, and don't start at the top of the list. The accompanying list of questions from the Contractor Reference Checklist can help you make the most of your phone interviews with a builder's previous customers.* If at all possible, personally visit two or three projects the contractor has completed. The more effort you invest at this stage of the project, the more likely it is that your construction project will be a success.

MEETING THE JOB SUPERVISOR

One of the most important questions you will have asked in the interview is whether the contractor will personally manage the project, or whether it will be managed by a job supervisor. As the Landrums learned so painfully, the success or failure of your project depends primarily on the person who manages the day-to-day construction.

If the contractor will not be personally managing your job, you need to meet the supervisor before deciding whether to hire the general contractor. It doesn't matter how impressed you are with the general contractor if you are not comfortable with the person you will be working with most of the time.

---

* *See Appendix A for a reproducible master of this form.*

Questions From Checklist 4.3

Contractor Reference Checklist

1. What kind of work did (insert contractor's name) do for you?
2. Was the work started on time?
3. Was most of the work completed on schedule?
4. Not counting requests for changes or extra work, was the work completed for the price quoted?
5. Do you feel the contractor handled changes and requests for extra work fairly?
6. Were you satisfied with the quality of construction?
7. How closely do you feel this contractor needs to be watched to make sure he/she is doing what he/she is supposed to do?
8. Was the contractor easy to contact when you needed to talk with him/her?
9. Were you satisfied with how the billing and paperwork were handled?
10. Did the contractor look after your interests during the project?
11. Was the contractor cooperative and easy to work with?
12. Were problems or misunderstandings resolved fairly?
13. Would you hire this builder again for the same project?
14. Would you recommend this builder to your best friend?
15. In hindsight, how could you have improved your working relationship with the contractor?
16. Do you have any other advice that would help me as I begin this process?

Thank the customer for taking time to answer your questions.

Deciding Which Contractors Make the Cut

The time has come to decide which contractors you want to invite to give you an estimate. Consider the following factors in making your decision:

✓ *Availability.* Occasionally a contractor you are interested in may not be available. Do not be too quick, though, to eliminate a good builder because he or she isn't available as soon as you would like. Keep in mind that most projects start later, some much later, than originally hoped because of such factors as design changes, planning and building department reviews, financing paperwork, and weather.

Rather than insisting on a specific start date, when talking to prospective contractors name a period of time during which you would like to start your project. Any contractor interested in your job will try to work out the schedule. Explore the possibility of adjusting your schedule or the contractor's or both so you can work together. If a builder's schedule is too full to take on your job on your timetable, respect that. An overcommitted builder is a ticket to disaster.

✓ *Specialization.* Does the builder specialize in your type of work? An example of specialization would be a kitchen cabinet shop that only does kitchens. Of course, a builder doesn't have to specialize in your type of work to be qualified to do it. The issue is whether the builder does your type of work regularly. For example, a custom home builder or remodeler who regularly works on kitchens as part of its larger projects would probably be well-qualified to remodel your kitchen.

- ✓ *Quality of work.* Based on your interview with the builder, interviews with at least six of the builder's previous customers, and personal inspection of the builder's projects, are you impressed with the builder's quality of work?

- ✓ *Business practices.* Does the company have the necessary license and insurance coverage? Is he or she prompt? Well-organized? Are you confident of this person's ability to handle the daily operation of a business with efficiency and excellence?

- ✓ *Compatibility.* Even if a builder passes all these tests with flying colors, that doesn't mean he or she is right for the job. What does your intuition tell you? Ignoring a negative hunch can amount to planting a time bomb that will not be revealed until it's too late.

If you are considering a specific builder but still have doubts or questions, discuss them with the builder. The builder may be able to resolve your questions simply by providing additional information.

Beware, though, of any concerns that you cannot trace to anything specific. Not being able to pinpoint what is causing your uneasiness can indicate a personality conflict that may lead to future problems. A good working relationship requires communication, cooperation, and trust. When considering a builder, ask yourself, "Do I want to work closely with this person on a day-to-day basis?" Unless your answer is a confident yes, eliminate that builder from your list. Keep interviewing contractors until you have found two or three that make the cut.

Only when all these steps are completed are you ready to apply the "three bids" rule.

Follow-up Calls

Call any builders you are not planning to use and thank them for their time. Call the two or three builders you want to consider further and ask if they are willing to prepare estimates or bids. If your final drawings are complete, you are ready to request final bids. If your drawings are at a preliminary stage, you will be asking for estimates. Chapters 5 through 7 will explain how to solicit and evaluate estimates and bids.

Chapter 4
At A Glance

✎ The purpose of the contractor interview is to learn as much as possible about the contractor's qualifications and personality.

✎ Preparing for the interview will increase its effectiveness. Arrange to prevent distractions. Know the questions you want to ask. Allow enough time for the interview.

✎ Remember: Just because someone is willing to give you a bid doesn't mean you should ask for one.

✎ After the interview, record your impressions. You may want to use form 4.2, "Contractor Evaluation," found in Appendix A.

✎ Make sure the contractor provides enough references. Talking to customer references is essential. Ask questions that require more than a yes or no answer, like those on the Contractor Reference Checklist (Checklist 4.3). Call references in random order. Call at least six, preferably more. Record their responses.

✎ If a general contractor does not plan to personally manage your project, meet the job supervisor who would manage it. The job supervisor is as important to your project's success as the general contractor.

- ✎ In deciding whom to ask for estimates, consider each builder's availability, specialization, quality of work, business practices, and compatibility with you.

- ✎ Keep interviewing builders until you have found two or three whom you feel you can work with and who you believe would do an excellent job. Only when all the steps to this point are completed are you ready to apply the "three bids" rule.

Chapter 5
Asking For Estimates

A younger couple had an older home in a good neighborhood. They didn't have a lot of money to invest, but wanted to make minor to moderate changes to all areas of the house.

They asked three builders for estimates. Since they knew exactly what they wanted done, they walked each builder through the entire house, pointing out the changes they wanted. Each builder took notes and gave them an estimate for the work.

Even before asking for estimates, the couple knew which builder they preferred, so they were disappointed when he turned in the highest bid. Feeling they could not afford the builder they really wanted, they hired another builder who had given a lower bid.

When the remodeling was finished, though, they had spent slightly more than their first choice builder had bid, and they were a bit disappointed in the quality of the work.

Lowballing

This couple fell victim to one of the oldest tricks in the business—lowballing. Builders, given vague information about a project, turn in unrealistically low estimates knowing they can make up the difference with change orders whenever the owners get around to deciding on specifics.

Getting estimates from honest but inexperienced builders can yield the same result. Because they may be less familiar with how much time and material a job will require, less experienced builders may turn in estimates that are too low. Then when homeowners choose a contractor, as this couple did, based on inaccurate estimates, disappointment almost always follows.

How can you avoid making the same mistake?

ESTIMATES VS. BIDS

First, don't confuse estimates and bids. While most customers—and many builders—use the terms *estimate* and *bid* interchangeably, knowing the difference between the two is critical to being able to hire the right contractor at the right price.

An **estimate** is *a calculation of the approximate cost to complete a general scope of work.* A **bid** is *an offer to perform certain work for a specified sum.* While an **estimate** gives you an approximate idea of what it will cost to build a project, a **bid** is a final price a builder submits for doing specific work based on final drawings and specifications.

The couple whose story opened this chapter chose a contractor based on preliminary estimates that had been calculated from incomplete information, not final drawings and specifications. The estimates they got were actually little more than guesstimates, practically useless for making a hiring decision. It would have been remarkable if their final bill had *not* come in considerably higher than the lowest bid.

Final bids, in fact, almost always come in higher than preliminary estimates, sometimes as much as fifty percent higher. The three most common reasons for this are:

(1) the preliminary specifications are incomplete and do not provide enough detail,

(2) the builder is inexperienced at estimating or fails to put in sufficient time to calculate an accurate estimate, and

(3) during the final design process, it is common for homeowners to request many small (and sometimes large) changes to the original design that, taken together, add substantially to the cost.

WHY GET PRELIMINARY ESTIMATES?

If preliminary estimates aren't precise enough to base contracts on, why even bother to get them? Why not just ask for final bids to start with? Because it is possible to request estimates in a way that insures reasonable accuracy, and when they are accurate, estimates can serve three useful purposes:

(1) Determining the approximate cost for the project. Early in the design process, before you invest a lot of time and money in detailed drawings, you want to know if you can afford to do everything you hope to do.

(2) Guiding budget-driven design decisions. If the estimates come in above your budget—a very common occurrence—they can guide you in whittling your plans down to what you are prepared to spend.

(3) Evaluating builders. Although estimates are by definition approximations, if you go about asking for estimates in the right way, the estimates you get can point you to the one or two builders from whom you should request final bids.

Be aware, though, that because of the time and cost involved, not all builders provide preliminary estimates. Most of these builders, though, will be glad to give you a firm bid when the final drawings are complete. You'll need to decide for yourself whether to invite such a builder to join the bidding process after you have received preliminary estimates from other builders.

How to Get Accurate Estimates

When one builder's estimate is half of another's, you can be pretty sure it is not because one builder can do the same job at half the price. The low-priced contractor may be lowballing or leaving out items the higher-priced contractor is including. Or the high-priced contractor may be busy and not really need the work. How can you insure that when you compare estimates, you are comparing apples with apples, not apples with pineapples?

1. ***Request estimates only from contractors whom you are seriously considering.*** When contractors know you are actively considering them for your job, they are willing to invest more time to prepare accurate estimates. Depending on the job, preparing an estimate can cost from a few hundred to several thousand dollars. Requesting estimates from contractors whom you know you will not hire is simply unethical.

   While you can always find people willing to bid your project, the best builders may ask you a couple of pointed questions before agreeing to work up an estimate:

   ✓ *Question #1: How many contractors are bidding the project?* The main reason builders ask this question is to get an idea of their chances of getting the job. The more builders bidding, the lower their chances. Bidding takes considerable time and effort, so contractors want to focus their limited resources on those projects most likely to lead to a contract.

Although quality builders are not necessarily the most expensive, they are seldom the cheapest. There will always be another contractor willing to do the job for less. The temptation to choose a contractor solely on price is stronger than most people realize or care to admit. Most contractors would agree that more than three bidders on a residential project indicates that price is the owner's primary consideration, and therefore it may not be worth a quality builder's effort to bid the project.

✓ *Question #2: Who are the other contractors and what is their background?* Professional builders know their prices will be higher than those of less scrupulous or less experienced contractors, so before they invest substantial time, they want to know they will be competing only with other quality builders. Builders who feel they have little chance of getting the job may not want to invest the time to prepare a bid.

2. ***Give the contractors specific information.*** Estimates can only be as accurate as the information they are based on. If drawings and specifications are vague, the estimates will be vague. While it is too soon to produce complete drawings and detailed specifications, the greater detail you can provide the builders, the more accurate their estimates will be.

Have your architect or designer provide as much of the following as you can:

Drawings

Basic drawings showing *existing* conditions

- ✓ Plot plan (drawing of property with buildings)
- ✓ Floor plan
- ✓ Exterior elevations (view of exterior of building)
- ✓ Roof plan
- ✓ One or two cross sections

Basic drawings showing proposed *changes*

- ✓ Plot plan (property with existing buildings and additions)
- ✓ Floor plan
- ✓ Exterior elevations
- ✓ Interior elevations (for areas such as kitchens and baths)
- ✓ Roof plan
- ✓ One or two cross sections

Specifications

As much as possible, list the brand, model, type, quality, material type, or other defining features of the equipment and materials your project will require. For example:

- ✓ Windows (wood, aluminum, or vinyl)
- ✓ Doors (interior and exterior, hollow or solid, stain or paint, grade)
- ✓ Cabinets (painted, stained, laminate, etc.)
- ✓ Appliances (quality example, brand or model)
- ✓ Door, cabinet, and bath hardware (brand and finish)
- ✓ Plumbing fixtures (brand, model, and finish)
- ✓ Moldings (paint or stain, size and type of material)
- ✓ Floor coverings (tile, carpet, wood, vinyl, etc.)
- ✓ Counter surfaces (tile, plastic laminate, wood, stone, etc.)
- ✓ Lighting fixtures (approximate budget per fixture)
- ✓ Interior wall finish (plaster or drywall, smooth or textured, etc.)
- ✓ Exterior wall finish (hardboard, wood, stucco, brick, stone, etc.)
- ✓ Roofing (tile, wood, concrete, metal, etc.)

Pictures

Photographs from magazines and product brochures can give builders a clearer idea of the level of quality you are looking for.

3. ***For any items with incomplete specifications, give builders a unit cost or allowance to include in their estimates or ask them to provide one.*** At the preliminary estimate stage of most projects, some items have not yet been selected. These unspecified items are where many of the price variations between bids and estimates arise. If one contractor can purchase a specified product or complete a clearly defined scope of work for less cost than another builder, this is a legitimate competitive advantage. However, when submitting estimates that include unspecified items, contractors may choose to include allowances that are too low and not realistic to make their estimates appear more competitive. When each contractor uses different allowances, the cumulative effect is that the estimates vary substantially. If, for example, Contractor A includes an allowance of $7,500 for cabinets and $25 per square yard for carpeting, while Contractor B includes $9,500 for cabinets and $29 per yard for carpet, a price difference of a few thousand dollars will result just between these two items.

   To minimize these confusing variations between estimates, contractors should be provided with a specific allowance for items with incomplete specifications. This allowance takes the form of a *unit cost*—for example, cost per square foot, square yard, or linear foot for such materials as moldings, tile, carpeting, and door and cabinet hardware. It takes the form of a *lump sum amount*—a total purchase price—to be budgeted for such purchases as cabinets, lighting fixtures, and appliances.

   You can do some shopping beforehand to get an idea of the price range for the quality of products you are looking for, or your architect can suggest allowances appropriate to your project.

4. ***Give each contractor the same information.*** A common reason bids vary so much is that homeowners, without realizing it, give each contractor different information. As the owners discuss the project with builders, each builder asks questions and new ideas come up. Each builder goes away with slightly different instructions. Preparing a well-defined package of written information to give each builder will reduce variations in estimates.

5. ***Set a specific due date for the estimate.*** Two to four weeks is enough for most single-residential projects.

6. ***Ask for itemized estimates.*** Without prices for specific components of the project—such as cabinets, framing, plumbing, or roofing—comparing estimates is almost impossible. Having prices for each item enables you to not only compare estimates, but to make cost-related design decisions.

7. ***Ask contractors to list any items not included in the estimate that may be required to complete the project.*** This encourages contractors to not leave things out of their estimates to appear more competitive. It also alerts you to budget for these expenses so they won't come as surprises later on.

8. ***Promptly inform all contractors of changes to the drawings or specifications.*** If some aspect of your plan changes while contractors are working on their estimates, call all partici-

pants and tell them of the changes. When you receive their estimates, check with the builders to make sure the estimates reflect the changes.

If the couple in the story at the beginning of this chapter had followed these steps, their choice of a contractor might not have been self-evident, but at least their three estimates would have all been for essentially the same services—comparing apples to apples. They would then have been ready to lay the estimates side by side, analyze each one, and determine which builder could give them the best value for their money. The next chapter explains how you can do that with the estimates you get.

CHAPTER 5
AT A GLANCE

✎ An **estimate** is *a calculation of the approximate cost to complete a general scope of work.* A **bid** is *an offer to perform certain work for a specified sum.*

✎ Three reasons final bids are usually higher than preliminary estimates: (1) information the homeowner gives builders is not specific enough; (2) the builder is inexperienced at estimating or puts too little effort into figuring the estimate; (3) the homeowner makes changes to the original design that increase costs.

✎ Three benefits of getting estimates: (1) to establish a project's approximate cost; (2) to guide the budget of the design process; (3) to evaluate potential builders.

✎ Not all builders provide preliminary estimates. Some prefer to bid on the final specifications only.

✎ Two questions good builders will probably ask you: (1) How many contractors are preparing estimates? (2) Who are they and what is their background?

✎ Eight steps to getting accurate estimates:

1. Request estimates only from contractors who have a fair chance of getting the job.

2. Give builders specific information.

3. For any items with incomplete specifications, have builders use the same allowances in their estimates.

4. Give each contractor the same information.

5. Set a due date for the estimate.

6. Ask for itemized estimates.

7. Ask the contractors to include any items not listed that may be required to complete the project.

8. Promptly inform all contractors of changes in drawings or specifications.

CHAPTER 6

COMPARING ESTIMATES

How to Decipher the Numbers

The Mitchells took great care in interviewing potential builders, selecting three to ask for estimates, and putting together their preliminary estimate package. All three builders submitted itemized estimates within 10% of each other, ranging from $97,000 to $106,000. But when the Mitchells laid out the three estimates side by side and tried to figure out why the estimates were different, they got frustrated. The three builders had all used different categories to break down their estimates, so the Mitchells had no way to compare the estimated cost of specific items. Had the contractor with the lowest estimate left out things the others had included? Was the one with the highest estimate really the most expensive, or just the most thorough? Though they spent several hours analyzing the estimates, they never were able to figure out the reasons for most of the $9,000 difference between the high and low bids.

WHY PRICES VARY

Estimates from different builders may vary up to twenty-five percent or even more. Bids so far apart are often not comparing apples to apples and may have little value to the homeowner. With careful planning, though, a homeowner can reduce the number of variables. When this is done, the resulting estimates will typically be much closer. What factors cause prices to vary, and how can homeowners reduce these variables so they can make meaningful price comparisons?

Factor #1: Drawings and specifications

Incomplete drawings and specifications force contractors to make assumptions about what will be needed or selected to complete the job. When required to guess, some contractors will lowball to make their estimates look attractive. Others will estimate high to be sure their bases are covered.

As the last chapter explained, there is a way around this. When the drawings specify items to be included in the estimate, but the specifications do not indicate brand, model, or other details necessary to calculate cost, the homeowner should specify an allowance to be used for each item. When this is done, the only cost variable that remains is how much the builder will charge to install the items.

Factor #2: Non-standard building methods

Difficult or unusual building techniques also invite wide variations in estimates. Labor often represents two-thirds of the cost of construction, so when builders don't know how much time something will take, they tend to err on the side of caution. Since builders interpret unfamiliar tasks differently, they also price them differently. When trying to minimize variations in estimates and hold costs down, avoid unusual or non-standard construction methods.

Factor #3: Supply and demand

Like any product or service, construction prices fluctuate depending on demand. When construction is booming, prices tend to be higher. During slow economic times, prices tend to be lower and more negotiable.

Demand varies with the seasons. During spring and summer months, builders are usually busy and their prices less negotiable. From late fall through early spring, their prices tend to become more flexible. To get the best price on your construction project, plan to ask for estimates when demand for construction is low.

Factor #4: The prices builders pay

Much of the contractor's price consists of smaller bids from suppliers and subcontractors. Subcontractors are companies that specialize in certain aspects of construction, such as plumbing, electrical systems, or roofing. Lower quality general contractors tend to choose subcontractors and suppliers that give them the lowest price. Better builders use subcontractors or suppliers who provide an appropriate balance of competence, quality, service, and price.

The builder's own employees may also do some of the work. The workers' level of skill and experience determines what the builder must pay them. Some managers hire the least expensive labor they can find while others hire only skilled workers. Better workmanship requires better carpenters. The cost and productivity of labor depends on which approach the contractor takes.

Factor #5: The builder

Differences in builders result in price differences.

✓ *Quality.* Builders known for high quality are more likely to include additional time in their estimates to allow workers to do a quality job. Quality-oriented builders also weigh how much supervision time they can include without pricing themselves out of a job. Less experienced builders rarely budget enough time for supervision.

✓ *Overhead.* Most smaller contractors have less overhead. Larger builders, though, must include allowances in their estimates for such overhead expenses as office space, equipment, office staff, and advertising. These supporting services provided by larger builders do not simply drive up costs; they contribute to the contractor's ability to provide quality and service.

✓ *Experience at estimating.* Accuracy in estimating comes with experience. Regardless of how long they have been in construction, builders with less experience in preparing estimates are more likely to:

✗ underestimate the time needed for a project,

✗ overlook or leave out some component of the process,

✗ fail to budget for unanticipated expenses,

✗ allow too little for insurance, labor-related taxes, overhead, and profit,

✗ or artificially lower the estimate to get the job.

✓ *Type of building experience.* Builders with different kinds of construction experience or different business philosophies will approach the job differently and so will submit different prices.

Factor #6: The customer

Just as each builder is different, so is each customer. Some owners have built or remodeled many projects while others are remodeling for the first time. Experienced builders will try to match the level of service to the customer. Some customers need more handholding than others. Some want extensive dust control measures, while others are less concerned about dust and simply want a quality job at the best price. Are builders overcharging when their bids include the cost of providing these services? No, they recognize that construction is more than boards and nails, and that to be successful they must meet or exceed the customers' expectations while at the same time earning a fair profit.

Contractors who fail to budget for such circumstances may become frustrated or angry when customers ask too many questions or make what the builder feels are excessive demands. They see dollar signs flashing by like the numbers on a gas pump and don't recognize that they should have more clearly understood the expectations of the client.

As you compare estimates, these six factors will account for most of the differences between estimates. Seeing behind the numbers to the reasons they are different is the key to being able to choose the builder who can give you the best value for your money.

HOW TO READ AN ESTIMATE

To accurately estimate the total cost of a job, a contractor first breaks the project down into many smaller components. He or she then calculates the labor, materials, and other costs required to complete each of these parts, then adds them to get the total cost.

As the sample estimate on the facing page shows, construction expenses are grouped first by division headings such as "concrete," "finish carpentry," or "plumbing." These divisions are further broken down into phases. In the example shown, electrical is subdivided into phases including wiring, service panel, lighting fixtures, and final installation. Phases may be further subdivided into items. For example, "framing" (under rough carpentry) can be further itemized to show the cost of framing by location.

For the purpose of comparing estimates, the level of detail shown in the example—divisions and phases—is most helpful. No greater level of detail is needed until the final bid stage.

Across the top of each page, the expenses are spread among five headings—labor, materials, subcontract, equipment, and other—with a column for totals at the right.

After adding all these subtotals together, the contractor adds allowances for other costs such as "overhead and profit" and "taxes and insurance" to arrive at the total project cost.

Why It's Hard to Compare

That all seems straightforward enough. So why did the Mitchells get so frustrated when they tried to compare estimates? Because no two contractors use exactly the same categories for calculating estimates. Even construction professionals find it difficult to compare two estimates and be able to tell whether one builder or the other is offering the better price on a specific item.

What is the solution? Ask the contractors to submit their estimates using some reasonably standardized format so you can compare estimates meaningfully. You may want to ask

Project Estimate

Smith Remodel
1234 Woodland Avenue
Any Town, USA

DESCRIPTION	LABOR	MATERIALS	SUBCONTRACT	EQUIPMENT	OTHER	TOTAL
Concrete						
Foundations			$2,500.00			$2,500.00
Concrete slabs			$67.00			$67.00
Subtotal	$0.00	$0.00	$2,567.00	$0.00	$0.00	$2,567.00
Rough Carpentry						
Framing	$5,500.00	$3,500.00				$9,000.00
Exterior windows & doors	$250.00	$7,500.00				$7,750.00
Subtotal	$5,750.00	$11,000.00	$0.00	$0.00	$0.00	$16,750.00
Finish Carpentry						
Moldings	$2,500.00	$1,700.00				$4,200.00
Cabinetry	$975.00	$7,500.00				$8,475.00
Interior doors	$550.00	$2,200.00				$2,750.00
Subtotal	$4,025.00	$11,400.00	$0.00	$0.00	$0.00	$15,425.00
Roofing						
Cedar shake			$11,000.00			$11,000.00
Subtotal	$0.00	$0.00	$11,000.00	$0.00	$0.00	$11,000.00
Plumbing						
Rough piping			$1,700.00			$1,700.00
Sewer line						$0.00
Plumbing fixtures		$2,200.00				$2,200.00
Fixture installation			$750.00			$750.00
Subtotal	$0.00	$2,200.00	$2,450.00	$0.00	$0.00	$4,650.00
Electrical						
Rough wiring			$1,400.00			$1,400.00
Service panel			$1,200.00			$1,200.00
Lighting fixtures		$1,750.00				$1,750.00
Final installation			$975.00			$975.00
Subtotal	$0.00	$1,750.00	$3,575.00	$0.00	$0.00	$5,325.00
Subtotals	**$10,300.00**	**$27,125.00**	**$27,500.00**	**$0.00**	**$0.00**	**$64,925.00**
Overhead & Profit						$9,700.00
Taxes & Insurance						$1,800.00
Total Project Cost						**$76,425.00**

builders to itemize their estimates using a form you provide, such as the Estimate/Bid Comparison Form (Form 6.1 in Appendix A).

If the Mitchells had asked the contractors to submit their estimates in this format, they could have more easily identified exactly how much the contractors' estimates varied on specific items. They would then have had a much better idea of why the amounts were different. Once they knew how much the prices differed on specific items and why, they could have more accurately determined which estimate represented the best value.

Of course, even if all the contractors who give you estimates are qualified to do the job, you don't base your choice of builder solely on dollars and cents. You still keep in mind what impressed you about these builders in the first place—their quality of work, their business practices, and their compatibility with you—and treat price as an important, but not the only, element in your decision.

Once you've added the price factor to your evaluation of the builders, you're ready to choose one or two builders as finalists for your job. Chapter 7 will tell you how to put together your final bid package, solicit one or more bids, and then evaluate those bids.

CHAPTER 6
AT A GLANCE

✎ Estimates can easily vary by up to 25% or more due to:

- ✓ Incomplete drawings or specifications
- ✓ Difficult or unusual construction techniques
- ✓ Supply and demand
- ✓ The prices builders pay
- ✓ Differences in builders
- ✓ Differences in customers

✎ When homeowners or their architects or designers carefully manage those variables in the estimating process that are within their control, the price variations between estimates are usually minimized.

✎ It is hard to compare the cost of specific items in estimates because no two builders itemize their estimates exactly the same way. Solution: Ask all contractors to provide you with price detail using the Contractor's Estimate/Bid Checklist (Checklist 6.1) as a guide.

✎ Don't choose a contractor on price alone. Consider price along with quality of work, business practices, reputation, and personal compatibility to select the one or two contractors you will ask for final bids.

CHAPTER 7

FINAL BIDS

Getting the Most for Your Construction Dollar

When Gene asked two contractors for final bids on remodeling his kitchen, he gave each of them a bid package consisting of complete drawings, a description of the work to be done, and many specifics of the job such as dimensions of the cabinets. Both builders bid the job tight, and Gene was pleased with the bid he accepted.

Once the work began, though, a lot of changes in the specifications were needed. Gene hadn't included specific brand names and make and model numbers for some of the items in his bid package, and he hadn't named specific patterns for tile and moldings. When Gene made his selections, most of them cost more than the builder had allowed. Not only did this increase the cost of materials, but the contractor added his standard percentage for overhead and profit for each of these change orders. While the contractor's additional charges were reasonable, Gene was well aware that he wasn't getting as good a price on these changes as he had gotten on the original bid.

Five years later, Gene added a family room and bath to his home. This time, because of what he had learned from remodeling his kitchen, Gene made his bid package as specific as possible. He even specified what kind of doorknobs he wanted. He listed the exact plumbing fixtures to be installed in the bathroom. He selected the brand and finish of bathroom cabinets he wanted. He had learned that every selection postponed until after the final bid leaves the door open for costs to keep increasing.

PREPARING FOR THE FINAL BID

Extremely important: The final bid is your single best opportunity to get the most value for your money, because this is when the contractor has the greatest motivation to submit the best possible price.

To take full advantage of this opportunity, you and your architect or building designer must prepare thoroughly for the final bid process. After receiving preliminary estimates and before requesting final bids, you have several important jobs to do.

1. ***Make final design decisions based on cost.*** Did the preliminary estimates come in above what you are willing to spend? If so, this is the time to trim your dreams to fit within your budget.

2. ***Have your architect or building designer complete the final drawings.*** At the preliminary estimate stage you used preliminary drawings since you didn't want to pay for final drawings in connection with work you might not end up doing. Now that you know what you can afford, you are ready to go ahead with final drawings.

3. ***Complete the specifications for your project in as much detail as possible.*** When you requested preliminary estimates, you probably hadn't yet chosen all your specific fixtures, appliances, finishes, moldings, etc. So long as the

contractors who were preparing estimates all used the same allowances for items with incomplete specifications, you avoided variations in the estimates due to these unknowns.

But now you are asking contractors not for ballpark estimates, but for specific dollar figures for which they are willing to do your project. For those dollar figures to be accurate, they must be based on detailed specifications. As Gene found, when they are not, the door is open for the final price to increase substantially beyond the accepted bid.

Is it possible to anticipate every decision you will need to make and include it in your specifications? Probably not. But the general rule is, the more specific you are, the fewer changes you will need to make later. This almost always saves you money.

HOW MANY BIDS SHOULD YOU GET?

Assuming you have gotten preliminary estimates from three or four builders, how many final bids should you get? How much do you narrow the field before taking this next step?

It may be obvious to you at this point that you do not want to further consider certain builders. Perhaps an estimate came in too high to be affordable, or so low as to seem unrealistic. Perhaps a builder showed a lack of professionalism in preparing the estimate, or maybe you picked up signs that you might have trouble working with a particular builder. For any of these reasons, or others, you may narrow your field.

After cutting such builders from your list, if you still have two or three builders you are actively considering, the most common approach is to ask each builder to prepare a final bid on your project. After receiving the final bids, you then choose which contractor you want for your job.

A less common but often effective approach is to tentatively select your contractor after you receive preliminary estimates, but before you complete your design work. This allows you to include your contractor in your design decisions and it offers you some important advantages.

INVOLVING YOUR CONTRACTOR IN DESIGN

When you can select your contractor early and include him or her in the design process, the contractor can often make suggestions that will save you time and money and improve the overall quality of the project. While a general contractor would not normally replace your architect or building designer (though some contractors are also building designers), your contractor can be a valuable player on your design team. Depending on your project, your design team may also include an interior designer/decorator, a kitchen designer, a lighting consultant, and/or a landscape contractor.

What happens, though, if your general contractor serves as a member of your design team, then turns in an unacceptable final bid? Are you stuck?

You don't need to be. An effective way to keep your options open is to agree with your builder on an hourly fee he or she will charge for participating in the design process. This assures the builder that he or she is not giving you valuable time for free, and also gives you the freedom to place the project out for competitive bids if you choose to do so. Contractors who take part in the preliminary design process will usually not charge you for the time if they are awarded the contract. If you and your builder agree to this approach, ask your builder to keep a simple log of time spent during the design phase. It won't hurt for you to do the same.

In most cases, though, when you choose a builder after getting preliminary estimates, you won't even need to get competitive final bids. The builder will prepare a final bid for you, and you and the builder will be able to negotiate a satisfactory contract.

How to Request Final Bids

The basic steps for requesting final bids are the same as those for requesting preliminary estimates, with one important difference: The specifications and drawings should be far more complete than they were at the preliminary estimate stage. To get the best possible price, your specifications and drawings should be as complete as possible.

When and if you asked for preliminary estimates, it was adequate for builders to break their estimates down into divisions and phases. To prepare final bids, they will probably price the job down to the item level, estimating the cost of every individual item needed to complete the job. Not only does this ensure the accuracy of the final price, but it is important to have prices for individual items in case you make any changes.

Although the contractor will probably complete the bid to the item level, this degree of detail is not normally provided to the homeowner unless the contractor is awarded the contract and the owner inquires about the price impact of possible changes. For example, if you decide to delete a garage window, the contractor will need to refer to the item price to know how much of the total window budget to give you credit for.

For a more complete description of the steps for soliciting bids, review the steps for requesting estimates in chapter 5. Here is a summary of those steps.

1. Prepare a bid package consisting of drawings of existing conditions, drawings of proposed changes, and specifications.

2. For any items with incomplete specifications, provide a unit cost or allowance to be used in calculating bids.

3. Give each bidder the same information.

4. Give the contractors two to four weeks to prepare their bids, and set a specific due date for the bids.

5. Ask contractors to list work or items not included in the bid that may be required to complete the project.

6. Promptly inform all bidders of any changes to the drawings or specifications. When you receive the bids, check with the builders to make sure the bids reflect the changes.

GOING OVER BIDS WITH BUILDERS

When a builder turns in a final bid, you will want to sit down with the builder and discuss it together. Here are some questions you should ask the contractor.

1. The plans and specifications will be included as part of any contract. Does your price include everything in the drawings and specifications?

2. What is not included in the price?

3. Will any additional work be required to complete the project?

4. When could you start the job?

5. Assuming no unusual circumstances, when can you guarantee completion?

6. What can be done to avoid cost overruns?

7. Did you find any of the specifications or details to be unclear?

8. What are your expectations of me as a customer?

9. How do you charge for extra work?

10. If I request changes for work, how will you price the work? What do you charge per hour for labor? Are rates different for different skill levels?

11. When are payments to be made and how much will each payment be?

12. If we have a misunderstanding, how will it be resolved?

Once you have gone over final bids with one or more builders, the time has come to make your final decision: Who do you want to build your project? Review everything you have learned including strength of references, experience, availability, personality, and price. Once you know the answer to that question, you are ready to move on to Step 8, negotiating the contract.

Chapter 7
At A Glance

✎ The final bid is the owner's single best chance to get the greatest value for the money spent.

✎ After you get preliminary estimates and before you request final bids, you have three tasks to complete:

(1) Make final design decisions based on cost.

(2) Have your architect or building designer complete the final drawings.

(3) Complete the specifications for your project in as much detail as possible.

✎ The most common approach is to ask two or three builders to prepare competitive final bids for your job.

✎ A creative alternative is to choose your contractor after the preliminary estimate and ask him or her to participate in your design team. To keep open the option of soliciting competitive bids, you can agree to pay the contractor an hourly fee for design consulting services should you end up awarding the contract to a different builder.

✎ Soliciting final bids is similar to requesting estimates with one important difference: Specifications and drawings for final bids should be far more specific than those used to generate estimates.

✎ In preparing final bids, builders will calculate the job cost in far greater detail than is normally done for preliminary estimates.

✎ When submitting bids, builders normally provide one total figure representing their price to complete the project according to specifications. Unless he or she is awarded the contract, a builder will not normally provide you with detailed pricing except when the information is needed when pricing requests for changes.

CHAPTER 8

CONTRACTS

The Key to Protecting Yourself

When Ken got home from work, he went as usual to check out the construction crew's progress on the new addition to his home. Among other things, the shower tile had been installed that day. He was surprised to discover that the tile only went six feet up the walls rather than all the way to the ceiling. He had always intended for the tile to completely cover the walls and assumed that was what the contractor would do.

Ken called his contractor and complained that the tile subcontractor hadn't done his job right. "Your contract calls for standard treatment of the shower tile," the contractor answered, "which is what we did."

"So far as I'm concerned, putting tile on the whole shower wall is standard," Ken said. "I would never consider doing it any other way."

"According to the building code," the contractor said, "six feet is standard height for shower tile. I can call the tile subcontractor back to change it, but we'll have to charge extra for the additional work."

When Renoyce, a single mother, hired a contractor to remodel her bathroom, she had to stretch her budget to the limit to come up with the money. When the workers pulled up the old floor tile, though, they uncovered a big problem. Water from the shower had been leaking onto the floor, evidently for several years, and the entire subfloor was rotten and had to be replaced. Worse, some of the water had seeped into an adjoining bedroom, and several thousand dollars worth of repair work was needed.

Renoyce felt betrayed. "You know I don't have that kind of money," she told the contractor. "You told me you could do the remodel for the contract amount, and now you're telling me you have to have several thousand more. What am I supposed to do?"

PREVENTING MISUNDERSTANDINGS, ANTICIPATING SURPRISES

While hiring a qualified builder is probably the single most important thing you can do to insure a smooth construction project, misunderstandings can develop with even the most reputable of builders. Misunderstandings arise when the contractor and homeowner bring different expectations to the project and they fail to spell out those expectations in the contract. Small misunderstandings, like the one between Ken and his builder, may cost a few hundred or a few thousand dollars; big misunderstandings can cost tens or even hundreds of thousands of dollars. A thorough contract will help to prevent these misunderstandings.

Renoyce's misunderstanding with her contractor was caused by a surprise—the discovery of hidden damage that had to be repaired.

While contractors cannot calculate in advance the cost of repairs they don't even know about, a good contract will specify who is responsible to pay for such "surprises" and how any charges for correcting them will be calculated. When this is done, surprises don't have to lead to misunderstandings.

An experienced contractor will normally include allowances in the bid for a few "minor surprises" such as having to move a small water pipe or an electrical outlet, and usually will not charge extra for these changes. However, the cost of correcting more substantial conditions not known to exist would not normally be included in the contract price.

The Purpose of Contracts

Contracts essentially define the rules of the game. No one would consider playing a game without knowing the rules. In much the same way, you should be cautious about entering into a contractual relationship unless the rules are clearly defined in advance. The contract spells out the responsibilities of each participant, what each party can and cannot do, and how any disputes will be resolved.

Disagreements are most likely to develop, not over major responsibilities, but over who is responsible for the little things. Since a construction project is made up of a lot of little things, these can add up to a lot of money.

If your contract doesn't clearly define who is to do what, you may end up hiring a lawyer or arbitrator to interpret the contract and make decisions that can have major financial consequences for you and your builder. And you may end up paying someone else to do work you thought was included in the original price.

What Goes into a Good Contract?

The typical construction contract is just one or two pages long and barely begins to describe the responsibilities of the signers. Don't settle for a typical contract. Tell your contractor that the contract will need to address the items on the Contract Checklist (Checklist 8.1 in Appendix A).

The length of the contract should reflect the size of the project. A thirty-page contract for a kitchen remodel would be overkill, while a two-page contract for building a new house is probably inadequate.

The contract needs to be written in plain, easy-to-understand language. When a contract is well-written, *anyone* reading the contract should be able to understand precisely what is and is not included, either because the contract states exactly what will be provided or the contract refers to drawings and specifications that state exactly what is provided.

Vague or incomplete specifications will be little help when your builder begins installing hollow core paint-grade doors when you were expecting solid stain-grade wood. Nor is a contract useful if it merely says the price includes materials and labor to install baseboards. The language in the contract or specifications must clearly indicate what is included: "Provide and install solid red oak baseboards pattern #475 to all rooms and closets, except bathrooms, garage, and laundry room." If you haven't yet chosen your baseboard, the specification might read: "Provide and install stain grade baseboards to all rooms and closets, except bathrooms, garage, and laundry room. A purchase allowance of $2.10 per lineal foot with tax is included."

This same level of detail needs to be included for each item in the project. Many drawings and specifications provided by architects and house designers include this information, but don't assume yours do. Check the drawings and ask yourself: "What kind of doors am I getting? Are the doorknobs brass or chrome? What about the hinges? Are the cabinets solid wood or veneer?" Answers to all these questions need to appear either in the drawings and specifications or in the contract. The contractor is responsible to provide you with a thorough list of what will and will not be provided if this has not been clearly defined in the drawings and specifications; it is your job to check the list to make sure it is complete.

Many builders use their own contracts that have been tailored to meet the needs of the particular job. Small builders usually have more general off-the-shelf contracts. Some construction-related organizations as well as the American Institute of Architects also provide standard contracts called boilerplate contracts. A boilerplate contract typically favors the builder or architect, but when modified can serve as an acceptable contract. While it is the contractor's responsibility to provide a contract, you will want to make sure the contract protects your interests. Some contractors may grumble about providing a detailed contract, but a little prudence now can prevent a lot of problems later.

EVALUATING CONTRACTS

Once your builder gives you a contract, read it with the following questions in mind:

- ✓ Do I understand what the contract says?
- ✓ Does the contract properly address all the items in the Contract Checklist?
- ✓ From reading the contract, is it clear to you what you are and are not getting?
- ✓ Am I willing and able to abide by what the contract requires of me—payment schedule, removing personal property, work to be performed by me, etc.?
- ✓ Acknowledging the needs of the other parties, am I reasonably comfortable with everything the contract says?

Remember that as a general rule the party with greater experience will draft a contract that favors its position. Because most homeowners don't have the expertise necessary to evaluate construction documents, you may want to consult an attorney before signing a contract.

The attorney you hire to review the contract should be experienced in construction law. Most contracts can be reviewed by an attorney for a modest cost. Whether the attorney's fee is a few hundred or a few thousand dollars, it is a good investment. Should misunderstandings arise, a well-written contract can save you many times the cost of the fee.

FEE STRUCTURE OPTIONS

What kind of fee structure is best for you? The answer depends on the nature and circumstances of your construction project. While other types of fee structures do exist, the following three are the most common, and in almost all cases, one of them should be well-suited to your job.

Fee Structure #1: The Fixed Bid

This is the most common approach. After reviewing the construction site, plans, and all specifications, the builder provides the owner with a specific dollar amount he or she will charge to complete a defined scope of work. The price usually includes all the costs involved in completing the work described in the drawings and specifications, plus taxes, insurance, overhead, contingencies, and profit.

Fee Structure #2: Time and Materials

The builder contracts for labor on an hourly basis. Costs for materials, subcontracted work, fees, etc., are paid by the contractor and billed to the owner with the builder adding a percentage fee to cover overhead and profit.

Fee Structure #3: Cost Plus a Fee

The contractor bills you his or her actual and direct costs incurred for labor and materials. In addition, you pay the contractor a predetermined flat fee or percentage of cost for overhead and profit.

CHOOSING THE BEST FEE STRUCTURE

What are the pros and cons of each of these fee structures? What kind of projects or circumstances is each best-suited for?

#1: The Fixed Bid

Benefits:

1. The price quoted should be the price you will pay; the contractor is responsible to complete the job for a specific price.

2. Billings and total payments are more predictable.

3. The owner has less need to monitor the contractor's costs.

Drawbacks:

1. The contractor receives the benefit of any savings in costs.

2. Without proper controls, costs for changes can be greater.

When most appropriate:

This traditional fee structure works well if your project drawings and specifications are clear, well defined, and complete.

#2: Time and Materials

Benefits:

1. The cost to perform small jobs can be less.

Drawbacks:

1. The cost to perform all but small jobs is usually higher.

2. The owner has a greater need to verify the contractor's costs.

3. The contractor benefits when the job cost increases.

4. This gives the contractor no incentive to complete the work quickly.

When most appropriate:

This fee structure should be used only in a few specific circumstances.

1. You can use this approach on small projects where being off an hour or two on the estimate can cause a 20% to 50% price variation. Examples: Hiring a carpenter to install a shelf or hang a

door; hiring a plumber to do a repair. If you insist on a fixed bid in such cases, the contractor will almost always bid higher than the actual cost of doing the work. Why? Because the contractor must base the bid on the maximum time the job could take.

2. You may have to use this approach when the work is so unusual, difficult, or undefined that it would be almost impossible to accurately estimate the time needed to complete the job.

3. You may want to consider this approach when the project consists of many smaller handyman-type jobs, when you are designing as you go, or when you are making frequent changes to a varied or smaller project.

#3: Cost Plus a Fee

Benefits:

1. It can save substantially on the cost of larger projects.

2. The contractor does not normally benefit when cost for the project increases.

3. Unless substantial changes are made, the owner has more flexibility to make changes and pay only the contractor's cost without incurring additional markup.

Drawbacks:

1. The owner assumes some risk for increased project cost.

2. The owner may need to take a more active role in reviewing subcontractor bids and material costs.

When most appropriate:

This contract is most commonly used for larger projects—large additions and remodels, new home construction, and commercial or industrial work.

This can work especially well when the owners already know which contractor they want to work with without feeling a need for competitive bidding. The contractor provides a fairly accurate estimate for the project and negotiates a specific fee for his or her services. That fee does not go up so long as the overall scope of the project does not change significantly.

Some changes can be made without affecting the contractor's fee. For example, changing from standard grade cabinets to top-of-the-line models would increase costs for materials and probably installation, but since it would not increase the contractor's supervisory time, the contractor's fee would probably not change. In this situation, the owner saves the added markup that the contractor would normally charge for a change order under a fixed bid contract.

Cost is also held down by getting competitive bids from subcontractors and suppliers. The homeowner and contractor decide together which bids to accept.

Negotiating a good contract takes considerable time and effort as well as some money for professional advice. It is worth the investment since the cost of a poorly drafted contract can be far greater—long delays in construction, thousands of dollars, and a lot of grief.

CHAPTER 8
AT A GLANCE

✎ The purpose of a contract is to clearly define in writing an agreement between two or more parties.

✎ A good contract prevents misunderstandings by spelling out in detail the expectations of both the homeowners and the contractor. This includes defining who is responsible for "surprises"—repairing hidden damage or dealing with unexpected subsurface conditions.

✎ You have the most bargaining power just before awarding a contract. That is when builders eager to get a new job are most willing to negotiate.

✎ Most construction contracts do not adequately describe the responsibilities of the parties involved. Be sure your contractor gives you a detailed contract that covers the relevant items in the Contract Checklist.

✎ In evaluating a proposed contract, seriously consider hiring an attorney with experience in construction law to review your contract. Work with your architect or designer to confirm that the contract addresses all the work described in your construction drawings and specifications.

✎ Choose the fee structure that is best for your project and your circumstances.

#1: The Fixed Bid works well on most projects when you want competitive bidding and when your drawings and specifications are clear, detailed, and complete.

#2: Time and Materials is best for (1) small jobs, (2) when the time needed for a project is almost impossible to accurately estimate, or (3) when you will be designing a small project as you go.

#3: Cost Plus a Fee can offer substantial savings and flexibility on large projects.

✎ Negotiating a good contract does require some effort, but the investment is small compared to the potential costs of a bad contract.

Chapter 9
How to Get the Most From Your Contractor

Nancy's friends had told her about their problems with contractors—schedule delays, cost overruns, unreturned phone calls. Determined not to let these things happen to her, she took exhaustive steps to locate, interview, and hire the best contractor for her project. From the moment construction began, she communicated often and thoroughly with her contractor, always staying on top of the work's progress. She regularly demanded answers from the contractor. When the schedule slipped, she reminded the contractor of the penalty for not finishing on schedule. Whenever she felt a worker or subcontractor was doing something wrong, she said so.

Yet for all her diligence, a key element was missing in Nancy's supervision—a team attitude. She based her relationship with her builder not on trust, but on intimidation. While the builder understood that Nancy was just trying to protect her own interests, her attitude actually made it harder for him to get his employees and subcontractors to do their best work. Why?

Because they would rush through their work so they could be gone before Nancy showed up. The contractor had to make an extra effort to make sure the workers did their jobs as well as usual. And while the remodeling did get done, getting there was unpleasant for everyone involved—the workers, the contractor, and the homeowner.

ALLIES OR ADVERSARIES?

More often than not, contractors and the homeowners who hire them view each other with some degree of caution, suspicion, or distrust. When trust is weak, every change, every problem, every conversation can be an occasion for suspicion. You and your contractor can invest more energy in attacking each other and defending yourselves than in completing the project. When this happens, not only do both you and your contractor endure unnecessary tension, but the quality and progress of the work usually suffer as well.

Just as a manager sets the tone in an office, you as the homeowner affect the mood of the construction team. As in any work setting, the working environment—whether it is confrontive and intimidating, or cooperative and supportive—directly impacts the quality and schedule of the project.

In a successful working relationship, the homeowner and contractor work together as partners. Like members of a football team who have different roles to play, they realize they are on the same side working toward a common goal.

How can you develop a relationship with your contractor that is primarily cooperative rather than confrontive? First, of course, you should hire a contractor you trust, one with whom you can communicate, one who has a history of good working relationships with customers. If you have conducted thorough interviews and reference checks, these are all things you should know about your contractor before you hire.

It's not enough, though, to hire a contractor with a team attitude; you must be a team player too. Here are seven ways you can build the kind of partnership with your builder that will make him or her *want* to work for you.

SEVEN KEYS TO EFFECTIVE TEAMWORK

Key #1: Get expectations out in the open.

Building trust between you and your contractor begins with your first contact. It continues as you interview, as you gather and evaluate estimates and bids, and as you negotiate the contract. A critical aspect of this process is for you and your contractor to recognize and accommodate each other's needs and expectations.

Your contractor brings some needs to the job that may be different from yours such as meeting payroll and making a profit to compensate for the risks of being in business. You bring such concerns as the quality of the work, whether the contractor will cut corners when you're not around, whether the job will be finished on time, and whether the workers are going to uncover any expensive surprises, such as dry rot, termites, or structural problems.

As you and your contractor listen to one another and try to be fair in responding to each other's needs, trust grows. The inevitable misunderstandings are more likely to be resolved fairly in this kind of relationship than in one of suspicion and mistrust.

Key #2: Keep communication clear and current.

Construction drawings alone don't tell your builder everything needed to complete the project. You will need to maintain regular and open communication with your contractor, and sometimes with other members of the construction team, on topics such as product information, installation procedures, requests for changes, or the quality of work being done. When your instructions are unclear or late in coming, work may have to be redone. This, of course, is time-consuming and expensive.

Make specific instructions involving additions, deletions, or modifications to the original drawings and contract in writing. When your contractor has questions for you, oral answers are usually fine, but to reduce misunderstandings you should keep a written record of your oral instructions and what was discussed. Not only can these records refresh your own memory of what was said when, but most people, including judges, find it hard to argue with well-organized, chronological notes.

Key #3: Plan ahead to prevent delays.

When most people think of construction, they think of delays. Yet every year, many thousands of projects both large and small are completed on schedule. What makes the difference between a project plagued with delays and one that is finished on time? While not all delay-causing factors can be controlled, most can. Here are some of the most common causes of delays and what you can do about them.

✗ *Owner inaction or indecision.* Some delays are caused when an owner puts off making product or material selections. If you have trouble making construction decisions, consider these approaches:

  ✓ Delegate the decision to someone you trust who has experience in the area involved (such an interior decorator).

- ✓ Get advice from key members of the construction team (architect, building designer, contractor, interior decorator, etc).

- ✓ List the pros and cons for each option. Consider not only the immediate cost, but also the long-term benefit of each option. Avoid options you are not really happy with. Saving a few dollars now is short-sighted if you'll be forced to live with a decision you'll regret.

✗ *Overcommitted contractors or building designers.* Because the building business is cyclical, construction professionals often have either too much work or not enough. When a contractor or building designer is working on too many projects, all the projects suffer. To keep this from hurting your project, take the following precautions when you hire building professionals:

- ✓ Ask how many projects they are working on.

- ✓ Ask how many projects they anticipate working on while working on yours.

- ✓ Ask when the project will be completed and how they plan to keep to that schedule.

- ✓ Consider adding a bonus or penalty clause to your contract to maintain accountability for completing the job on time. Check with your attorney to see if this is appropriate in your circumstances.

- ✓Put agreements about timeliness of performance in writing.

✗ *Planning and building department delays.* Planning and building departments are just as affected by the building cycle as contractors are, so permit approvals often take longer during the busy season. To keep your project on track, try the following:

- ✓ Ask the city how long departmental approval typically takes.
- ✓ Be sure to include all requested information the first time when submitting packages for approval.
- ✓ The person applying for the permit should personally visit the building department each week. This helps to keep your project from getting buried on somebody's desk.
- ✓ Be polite to building department officials, even when you are frustrated. You will win more cooperation with courtesy than with anger.

✗ *Financing and payment delays.* As with getting permits, polite persistence can expedite approval of your construction loan. In addition, to keep construction moving, make sure you make payments to your contractor on time. Withholding money for insignificant items is counterproductive. People who pay their bills promptly get better service and gain more cooperation.

✗ *Back ordered or unavailable materials.* Order materials far enough in advance to avoid delays. Expect some items to arrive incorrect or damaged and allow plenty of time to resolve such problems. If delivery time for an item is excessive, choose a different item or ask about special shipping options. What seems like a lot of money for quicker freight may be insignificant compared to the cost of delaying the whole project. Check regularly on the status of items on order.

✗ *Conditions discovered after work begins.* Abandoned septic tanks, inadequate wiring or plumbing, asbestos, dry rot, and termite damage are a few of the many problems workers may discover once work begins. In most cases your contractor should quote a specific price to repair the problem condition. If the builder is not sure of the extent of the work, consider authorizing part of the work for an agreed amount so the builder can keep working until an exact quote is possible.

✗ *Excessive changes.* Making sure you are pleased with the finished product may occasionally mean changing work already completed. On the other hand, making lots of changes because you can't make up your mind lowers worker morale and productivity.

✗ *Weather.* While you cannot control the weather, you can take it into account in scheduling your project. Be realistic. Assume some delays will occur.

Key #4: Resolve misunderstandings.

A misunderstanding is not an intentional act, but a breakdown in communication. Some misunderstandings are resolved easily when one person suddenly realizes what the other is trying to say. Others take more work.

Most misunderstandings over what is or is not included in the contract can be prevented by a carefully detailed contract. But even with a good contract, some misunderstandings will arise. When they do, how can they be cleared up? The following guidelines will help. *(Note: This information is not provided as legal counsel or as the only means for resolving conflicts.)*

✗ *Ask, "Is it standard practice to include this item in the contract?"* "Standard practices" vary and many homeowners won't know what items usually are or are not included in construction bids, but can find out by checking with other contractors. Examples: Costs of permits are often not included in the bid unless specifically mentioned in the contract. On the other hand, a contractor who specialized in bathroom remodeling probably should allow for a few hours of work to correct *minor* (not major) hidden damage, such as water damage.

✗ *If the contractor failed to include necessary work in the estimate due to an inadequate review of the site and plans and specifications, the contractor would normally be responsible to include the work.* At this stage in the project, this should occur only with smaller items since major oversights would probably have surfaced during the review of the bid.

- ✗ *If you and your contractor are at an impasse, the two of you might agree to split the cost of the additional work based on the contractor's wholesale or actual cost without any charge for overhead or profit.* This may be especially appropriate in situations where an outside observer might say that both parties are partly correct.

- ✗ *To get an objective viewpoint, talk with your architect, building designer, or other contractors.* An outsider may be able to help one or both parties better understand the other's point of view.

- ✗ *If all else fails, you may need to hire an attorney or arbitrator with experience in construction law.* Your contract should say whether disputes will be handled by litigation or arbitration. Arbitration is the more common choice as it reduces legal fees.

Key #5: Visit the job site regularly.

Visiting the project regularly, daily if possible, lets the workers and the contractor know that you are willing to take an active role and will be available to answer questions. And with regular visits, you can discover mistakes sooner, reducing the cost of correcting those mistakes.

Most contractors are pretty busy first thing in the morning as they get the job started and plan for the day. Certain times of the day are better than others to have some uninterrupted time with the builder. Check with your builder about what time of day works best.

As you think of questions you want to ask your contractor, write them down. Not only will this help you to remember all the questions you need to ask on your next visit to the site, but it will help you make more efficient use of your time with the contractor.

Most construction sites are busy and potentially dangerous. Dress appropriately. If you will be wearing good clothes when stopping by the job site, at least keep a pair of tennis shoes and a light jacket in your car to lessen the chance of damaging your clothing.

Key #6: Expect an emotional roller coaster.

Seeing your home constantly torn up and dusty and having strangers in your home takes its toll. Give yourself, your spouse, and your children a little more slack during this time. More than one marriage has ended over the stress associated with construction. Whether yours is a six-week remodel or an eighteen-month custom home, remind yourself that this too will pass, and in the meantime, your family is more important than your project.

You will almost certainly experience a predictable cycle of emotional highs and lows during construction. Expect it.

In the beginning as demolition and framing get underway, the excitement of getting started usually overshadows any problems or early inconveniences.

Later, during the installation of rough plumbing, electrical wiring, and the ducting system, many owners feel the project is dragging. During this stage you will be busy answering the contractor's questions: *Where do you want the light? How high should the shower head be? What type of appliance will you be using?* You will also be visiting stores to make final product selections.

When the drywall is finally installed, most customers feel a sense of accomplishment and relief. It looks like it's almost finished! In reality, though, the project is only about half done. This is because the remaining work is slower and more detailed.

Installing cabinets, doors, moldings, tile, counters, lighting fixtures, plumbing fixtures, finish hardware, and so on, must be done in the proper sequence and it takes time to do these jobs right. During this stage, homeowners sometimes feel that not much work is being done and that what is being done is taking forever.

Finally, most owners try to move in too soon. When they do, this makes the last five to ten percent of the project take twice as long as it should because workers have to work around the furniture and the family. If possible, wait until the work is finished to move in.

Key #7: Learn the language.

Learning the terms used in construction will help to minimize misunderstandings between you and your contractor. Words used in this book with which you may not be familiar are probably defined in Appendix B: "Explanation of Terms." Your library or local bookstore has books that explain the many technical construction terms you are likely to hear while work is in progress. Take time to become familiar with those that relate to your project.

If you don't understand the meaning of a term your contractor is using, ask. Don't be embarrassed that you know less about construction than your contractor; that's why you hired a construction expert.

Do People Enjoy Working for You?

What type of boss do you like to work for? One who constantly criticizes your work? One who says you're not working fast enough? One who constantly tells you how to do your job? Or would you rather work for someone who respects you and your work, and who offers a little praise and encouragement from time to time?

Be friendly, respect your contractor's need to make a reasonable profit, provide information when it's needed, pay your bills promptly, be generous with sincere compliments—and an occasional blueberry muffin in the morning never hurts. Being reasonable doesn't mean avoiding conflict. You will at times need to hold your contractor accountable. But by being the kind of homeowner builders enjoy working for, you are almost sure to have a more effective partnership with your contractor, and you, your contractor, and your project will all come out ahead.

Chapter 9
At A Glance

✎ In a successful working relationship, the homeowner and contractor relate as partners, not adversaries.

✎ Seven keys to effective teamwork:

- ✓ Recognize and address each others' expectations.
- ✓ Keep communication clear and current.
- ✓ Plan ahead to prevent delays.
- ✓ Resolve misunderstandings.
- ✓ Visit the job site regularly.
- ✓ Expect an emotional roller coaster.
- ✓ Learn the language.

✎ Be the kind of homeowner contractors enjoy working for and you'll get the most for your construction dollar. This will almost surely lead to a productive working partnership in which you, your contractor, and your project all win.

Postscript

To some homeowners and contractors, these nine steps might seem rigorous. Yes, they are tough. They are tough because the time to find out if your builder is qualified and will remain committed to your project is before he or she starts.

Certainly no perfect builder exists, and even good builders make mistakes. Despite your best efforts to find the most qualified builder and to maintain excellent communication, misunderstandings will occur. With this in mind, don't discount a terrific candidate just because he or she doesn't meet all your expectations or doesn't fully measure up to every ideal mentioned in this book. Rather, use the nine steps as tools to help you make informed choices.

Transforming your hopes and dreams from ideas and lines on paper to reality that becomes a part of your everyday life is exciting. Keep your eye on the goal, but enjoy the journey too. The temporary dust and inconvenience will soon be forgotten, while the benefits of a well-designed and well-built home will be yours to enjoy for years to come.

Appendix A

Forms and Checklists

Note: Form numbers relate to the point in the nine-step process when the forms will be used. Form 4.2, for example, is to be used with chapter 4 and is the second form used in chapter 4.

Using Forms to Make Better Decisions

A well-designed form is a valuable tool. Each form in this section is designed to make it easier for you to gather, organize, evaluate, and reference the information you need to make wise decisions about your home construction or remodeling project. Not only can using systems save you time and frustration, they can save you money and help you avoid major grief.

Form 2.1

Contractor List

Questions to ask when gathering names of contractors:

Do you know of any contractors who specialize in ______________ (your type of project)?

Have you worked with this builder? How many times?

Were you satisfied with the quality of the contractor's work and service?

Would you hire this builder again?

Do you know of anyone who has recently done a similar project?

List any contractors you want to check out.

	Company or Name	License #	Phone #	Referred by
1.	____________	____________	____________	________
2.	____________	____________	____________	________
3.	____________	____________	____________	________
4.	____________	____________	____________	________
5.	____________	____________	____________	________
6.	____________	____________	____________	________

7. ____________________ ____________________ ____________________ __________

8. ____________________ ____________________ ____________________ __________

9. ____________________ ____________________ ____________________ __________

10. ____________________ ____________________ ____________________ __________

11. ____________________ ____________________ ____________________ __________

12. ____________________ ____________________ ____________________ __________

13. ____________________ ____________________ ____________________ __________

14. ____________________ ____________________ ____________________ __________

15. ____________________ ____________________ ____________________ __________

16. ____________________ ____________________ ____________________ __________

17. ____________________ ____________________ ____________________ __________

18. ____________________ ____________________ ____________________ __________

19. ____________________ ____________________ ____________________ __________

20. ____________________ ____________________ ____________________ __________

CHECKLIST 2.2

PEOPLE WHO MAY KNOW CONTRACTORS

1. List the names and phone numbers of people who may know of a contractor. Check off only after you have talked with them.
2. If they don't know of any builders, ask if they know of someone who does.
3. Have the Contractor List, Form 2.1, handy when talking with people. Use the questions at the top of the page as a reference.
4. Asking more people improves your chances of finding good builders.

People you know

WHO	***NAME***	***PHONE NUMBER***
( ) Accountant	______________________	____________
( ) Banker	______________________	____________
( ) Boss	______________________	____________
( ) Coworker	______________________	____________
( ) Dentist	______________________	____________
( ) Doctor	______________________	____________
( ) Friend	______________________	____________
( ) Friend	______________________	____________
( ) Insurance agent	______________________	____________

( ) Neighbor ______________________ ____________

( ) Neighbor ______________________ ____________

( ) Real estate agent ______________________ ____________

( ) Real estate agent ______________________ ____________

( ) Relative ______________________ ____________

( ) Relative ______________________ ____________

( ) Relative ______________________ ____________

People who work with contractors

WHO	NAME	PHONE NUMBER
( ) Architect	______________________	____________
( ) Architect	______________________	____________
( ) Banker (local)	______________________	____________
( ) Building department	______________________	____________
( ) Building designer	______________________	____________
( ) Building inspector	______________________	____________
( ) Cabinetmaker	______________________	____________
( ) Consultant	______________________	____________
( ) Electrician	______________________	____________
( ) Fine quality builders hardware store	______________________	____________
( ) Interior decorator	______________________	____________

( ) Lumber yard ______________________ ____________

( ) Plumber ______________________ ____________

( ) Plumbing store ______________________ ____________

( ) Real estate agent ______________________ ____________

( ) Structural engineer ____________________ ____________

( ) Tile Shop ______________________ ____________

( ) Window supplier ______________________ ____________

Professional Organizations

ASSOCIATION ***PHONE NUMBER***

( ) NAHB National Association of Home Builders ____________

( ) BIA Building Industry Association ____________

( ) NARI National Association of Remodeling Ind. ____________

( ) ABC Associated Builders and Contractors ____________

( ) AGC Associated General Contractors ____________

( ) BBB Better Business Bureau ____________

( ) AIA American Institute of Architecture ____________

( ) ASID American Society of Interior Designers ____________

( ) Board of Realtors ____________

( ) Builders exchanges ____________

( ) Chamber of Commerce ____________

FORM 3.1

CONTRACTOR TELEPHONE INTERVIEW

Company name ______________________ Phone ______________

Person ______________________________ Date ______________

Address ____________________________ License # __________

____

Recommended by ____________________

STEP 1: BACKGROUND CHECK

Contact these agencies and ask the following questions.

Better Business Bureau

✎ Number of years in business __________

✎ Have any complaints been filed? ( ) yes ( ) no

✎ How many? When? ____________________________

✎ Were the complaints resolved? ( ) yes ( ) no

Your state agency for handling consumer complaints *(such as attorney general)*

✎ Have any complaints been filed? ( ) yes ( ) no

✎ How many? When? ____________________________

✎ Were the complaints resolved? ( ) yes ( ) no

Local building department

- ✎ Has this builder completed any projects in town? ( ) yes ( ) no
- ✎ Do you work with this builder often? ( ) yes ( ) no
- ✎ Have you had any complaints against this builder? ( ) yes ( ) no

State licensing board for contractors *(if your state requires contractors to be licensed)*

- ✎ Is this builder's license current? ( ) yes ( ) no
- ✎ Is the license issued to the person or company I am considering hiring? ( ) yes ( ) no
- ✎ Does the contractor have a current license bond in good standing? ( ) yes ( ) no
- ✎ Have there been any complaints against this license? ( ) yes ( ) no

STEP 2: TELEPHONE PRESCREENING

Conduct a preliminary phone interview with the contractor.

It helps to qualify yourself if you tell the contractor who gave you his or her name. Then ask:

1. What type of construction do you specialize in? (If possible, let the contractor describe what kind of work he or she specializes in before describing your project.) ____________________

   ____

   ____

2. What kind of projects are you working on now? __________

   ____

   ____

3. What distinguishes your company from other builders? ________

   ____

4. How long have you been in business? ________________

5. Is your office nearby? ________________________

6. Do you have any completed projects in the area? __________

   ____

7. When can you start new projects? __________________

8. What do you think I should look for when hiring a contractor?

   ____

9. Do you carry workers' compensation insurance? ( ) yes ( ) no

   Liability insurance? ( ) yes ( ) no

Following the phone interview:

Do you want to schedule a personal interview with this contractor?
( ) yes ( ) no

If so, be sure to schedule the meeting so everyone who will be involved in selecting the contractor can attend.

Date ________________________

Time ________________________

Place ________________________

Form 4.1

Contractor Personal Interview

Company name ______________________ Phone ______________

Person ______________________ Date ______________

Address ______________________ License # ___________

____

Recommended by ______________________

PRELIMINARY QUESTIONS

1. Can you tell me about your background and how you got into construction? ______________________

 ____

2. How long have you worked in the construction field? ___________

3. What do you like the most and least about being a general contractor? ______________________

 ____

4. How much time do you spend on the job and what is your role in the construction? ______________________

 ____

5. What distinguishes your company from other contractors that provide similar service? ______________________

 ____

 ____

PRIMARY QUESTIONS

1. Do you like to become involved in the construction design process? Why? ______________________________

____

____

2. What are the keys to a successful construction project? ________

____

____

3. What are some common problems contractors face when working with owners? ______________________________

____

____

4. What are some common problems owners face when dealing with contractors? ______________________________

____

____

5. How can we avoid these problems? ______________________________

____

____

6. How do you ensure that the project is completed on time? ______

____

____

7. What can we do to keep the final bids within our construction budget?______________________________

____

____

8. Once we start the project, how do you ensure that our project stays within budget? ______________________________

____

____

9. How do you price requests for changes or additional work? ______

____

____

10. How do you handle complaints or misunderstandings during the construction project? ______________________________

____

11. Do you subcontract any of the work? Why? ______________

____

____

12. Who supervises the work? ______________________

____

____

13. How long have you worked with the subcontractors you use? ____

____

____

14. How do you keep your contractors accountable to the schedule and quality standards? ______________________________

____

15. How do you ensure the quality of your work? ______________

____

16. How is your payment policy structured? __________________

____

____

17. How can we be sure the general contractor is paying the subcontractors? ____

____

18. Under what circumstances should we withhold payment? ______

____

CLOSING QUESTIONS

1. Do you have any photos of previous projects we can see? ______

____

2. Do you have a list of references you can leave with us? ______

____

3. Can we see some of the projects you have completed? ______

____

4. Do you have any projects currently under construction that we could visit? ____

____

5. Do you have a sample of your contract with you? ( ) yes ( ) no

6. Do you provide preliminary estimates? ( ) yes ( ) no

7. How available are you when we need to contact you during the project? ____

8. How much variance is there between your preliminary estimates and your final bids? ____

____

9. What advice can you give us on how to choose a contractor? ____

____

____

FORM 4.2

CONTRACTOR EVALUATION

Use this tool to evaluate each contractor you interview in person. After the personal interview fill out Part I. After calling the contractor's references fill out Part II. Finally, fill out Part III. Circle the number that goes with the answer that is most nearly true for each question.

Contractor's name ______________________________

Date of personal interview ______________________________

Names of references checked ______________________________

____

____

____

Part I: Interview

THIS CONTRACTOR:	FALSE	UNSURE	TRUE
1. Arrived on time for the appointment.	0	1	3
2. Was well-prepared for the meeting.	0	1	3
3. Listened and responded to what I said.	0	1	3
4. Gave satisfactory answers to my questions.	0	1	3
5. Seemed to understand my ideas.	0	1	3
6. Acted interested in my project.	0	1	3
7. Has completed similar projects in the area.	0	1	3

8. Brought the information I requested.	0	1	3	
9. Never became impatient or irritated.	0	1	3	
10. Avoided criticizing previous customers, employees, and other firms.	0	1	3	

Part II: References

Talk with at least six of the contractor's references before answering this question.

	No	*Somewhat*	*Yes*	*Very*
11. Are this contractor's previous customers satisfied with the quality of work and service they received?	0	10	20	30

Part III: Gut Feelings

	No	*Somewhat*	*Yes*	*Very*
12. Do you think you would be comfortable working with this person on a daily basis?	0	5	15	20
13. Do you have a favorable overall impression of this builder?	0	5	15	20

TOTAL OF CIRCLED NUMBERS (100 possible points): [______]

Note: This tool is one guide to help you evaluate contractors. Factors not listed here may be relevant to your decision, and you may wish to give some factors more or less weight than they are given here.

CHECKLIST 4.3

CONTRACTOR REFERENCE CHECKLIST

After conducting personal interviews, call at least six of the previous customers of each contractor you are seriously considering. Begin with the middle or end of the contractor's reference list and call in random order. Record customers' responses on this form.

Name of contractor ____________________

Name of customer reference ________________

Customer phone number ____________________ Date interviewed ______

Location of project ____

1. What kind of work did (insert contractor's name) do for you? _____
   ____
   ____
   ____
2. Was the work started on time? ( ) yes ( ) no
3. Was most of the work completed on schedule? ( ) yes ( ) no
4. Not counting requests for changes or extra work, was the work completed for the price quoted? ( ) yes ( ) no
5. Do you feel the contractor handled changes and requests for extra work fairly? ( ) yes ( ) no
6. Were you satisfied with the quality of construction? ( ) yes ( ) no

7. How closely do you feel this contractor needs to be watched to make sure he/she is doing what he/she is supposed to do? ______

   ____

   ____

8. Was the contractor easy to contact when you needed to talk with him/her? ( ) yes ( ) no

9. Were you satisfied with how the billing and paperwork were handled? ( ) yes ( ) no

10. Did the contractor look after your interests during the project? ( ) yes ( ) no

11. Was the contractor cooperative and easy to work with? ( ) yes ( ) no

12. Were problems or misunderstandings resolved fairly? ( ) yes ( ) no

13. Would you hire this builder again for the same project? ( ) yes ( ) no

14. Would you recommend this builder to your best friend? ( ) yes ( ) no

15. In hindsight, how could you have improved your working relationship with the contractor? ____________________

    ____

    ____

16. Do you have any other advice that would help me as I begin this process? ____________________

    ____

    ____

Thank the customer for taking time to answer your questions.

FORM 6.1

ESTIMATE/BID COMPARISON FORM

(Check or indicate price for all that apply.)

	CONTRACTOR 1			CONTRACTOR 2			CONTRACTOR 3		
Description	Labor	Material	Allowance if applicable	Labor	Material	Allowance if applicable	Labor	Material	Allowance if applicable
Architectural									
Building design and plans									
Surveying									
Engineering									
Permit processing									
Other									
Job overhead									
Bond									
Permit fee									
Utility service fee									
Builders risk insurance									
Use tax									
Site security									
Temp utilities setup									
Portable toilet									
Job trailer									
Job phone									
Ongoing trash removal									
Other									
Site work									
Lot clearing/demolition									
Debris removal									
Grading/excavation									
Retaining walls									
Soils removal									
Erosion control									
Site drainage									
Patios and walks									
Decks/railings									
Drives									
Curbs and gutters									
Landscaping									
Finish grading									
Fencing									
Other									

Description	Contractor 1			Contractor 2			Contractor 3		
	Labor	Material	Allowance if applicable	Labor	Material	Allowance if applicable	Labor	Material	Allowance if applicable
Foundations									
Excavation									
Forming									
Steel									
Hardware									
Block work									
Concrete									
Backfill									
Slabs									
Other									
Rough carpentry									
Framing									
Facias, siding, and trim									
Windows									
Exterior doors									
Garage doors									
Other									
Roofing									
Elevated									
Flat work									
Flashing									
Gutters and downspouts									
Skylights									
Other									
Electrical									
Rough									
Main panel									
Subpanel									
Finish trimout									
Fixtures									
Alarms									
Other									
Plumbing									
Rough-in									
Trimout									
Sewer connection									
Septic systems									
Water main									
Fixtures									
Fire sprinklers									
Other									

Description	Contractor 1 Labor	Contractor 1 Material	Contractor 1 Allowance if applicable	Contractor 2 Labor	Contractor 2 Material	Contractor 2 Allowance if applicable	Contractor 3 Labor	Contractor 3 Material	Contractor 3 Allowance if applicable
HVAC									
Ducting									
Furnace									
A/C									
Exhaust fans									
Baths									
Kitchen									
Laundry									
Prefab fire boxes & chimneys									
Other									
Insulation									
Floors									
Walls									
Ceilings									
Other									
Drywall									
Hang									
Tape and finish									
Finish type									
Other									
Finish carpentry									
Doors									
Cabinets									
Door casing									
Window casing									
Baseboards									
Crown moldings									
Mantles									
Paneling									
Soffits									
Shelving									
Closet shelves and rods									
Stairs and railings									
Other									
Masonry									
Interior veneer									
Exterior veneer									
Fireplaces & chimneys									
Concrete block									
Precast concrete products									
Other									

Description	Contractor 1 Labor	Contractor 1 Material	Contractor 1 Allowance if applicable	Contractor 2 Labor	Contractor 2 Material	Contractor 2 Allowance if applicable	Contractor 3 Labor	Contractor 3 Material	Contractor 3 Allowance if applicable
Lath & plaster									
Interior plaster									
Exterior stucco									
Other									
Hardware									
Structural									
Entry									
Cabinet									
Doors									
Other									
Ornamental Iron									
Interior railings									
Exterior railings									
Decorative									
Other									
Waterproofing									
Showers									
Decks									
Caulking									
Building paper									
Flashings									
Other									
Tile									
Floors									
Tub and shower surrounds									
Counters									
Fireplace surrounds									
Other									
Glazing									
Mirrors									
Tub/shower enclosures									
Fixed glass									
Shelving									
Other									
Solid surfaces									
Fireplace surrounds									
Counters									
Tub/shower surrounds									
Shelving									
Other									

Description	Contractor 1 Labor	Contractor 1 Material	Contractor 1 Allowance if applicable	Contractor 2 Labor	Contractor 2 Material	Contractor 2 Allowance if applicable	Contractor 3 Labor	Contractor 3 Material	Contractor 3 Allowance if applicable
Floors									
Wood									
Carpeting									
Sheet vinyl									
Vinyl tile									
Other									
Equipment									
Exhaust hoods									
Appliances									
Central vacuum									
Elevators									
Other									
Painting & decorating									
Paint interior									
Walls/ceilings									
Doors & trim									
Cabinets/shelving									
Paint exterior									
Walls									
Facias, eaves, trim & gutter									
Doors									
Decks and rails									
Wallpapering									
Window coverings									
Other									
Other									
Supervision									
Jobsite superintendent									
Contingency									
Overhead and profit									
Total estimate									

CHECKLIST 8.1

CONTRACT CHECKLIST

Check over the draft of your contract to make sure it includes all the following elements.

Agreements

✎ Details of any prior or verbal mutual agreements

Allowances

✎ Unit costs for unspecified material such as moldings, carpeting, tile, etc.

✎ Total allowances for unspecified items such as light fixtures, appliances, cabinets, etc.

✎ Whether sales tax is included in the prices

Change orders

✎ Procedures for making changes

✎ How the contractor charges for change orders

Clean up

✎ How often

✎ How complete

Contractor information

✎ Name

✎ Physical home or office address

✎ Phone number

Customer information

- ✎ Name
- ✎ Address
- ✎ Job address
- ✎ Telephone number

Date

- ✎ The day the contract is signed

Documents

- ✎ Copies of and reference to drawings and specifications
- ✎ Instructions to bidders

Exclusions

- ✎ Work or items that will not be provided
- ✎ Who will pay for hidden or subsurface conditions

General conditions

- ✎ Conditions or rules by which the construction process will be conducted
- ✎ Standard "boilerplate contract conditions" may vary from state to state

Indemnification

- ✎ Conditions that protect the owner's interest

Insurance

- ✎ Workers' compensation (contractor)

  Company

  Policy number

  Expiration date

- ✎ Liability insurance (contractor)

  Company

  Policy number

  Expiration date

- ✎ Automobile insurance (contractor)

  Company

  Policy number

  Expiration date

- ✎ Third party liability (owner)

  Company

  Policy

  Expiration date

License information

- ✎ Number
- ✎ Name of person to whom license is issued if different
- ✎ Address to whom it is issued
- ✎ Bond company (if applicable)
- ✎ Classification
- ✎ Expiration date

Lien releases

- ✎ Requirement that contractor will provide lien releases from sub-contractors and material suppliers
- ✎ Requirement that contractor will provide lien releases for all payments received

Materials

- ✎ Description or identification of materials to be used
- ✎ Brands, colors, model numbers, or other specifically identifying information
- ✎ Misunderstandings and how misunderstandings will be resolved.

Payment schedule

- ✎ Deposit amounts
- ✎ Sequential payment amounts
- ✎ Description of completed work to which payments will be connected
- ✎ Circumstances under which payments may be withheld
- ✎ Conditions under which final payment is due

Penalty clause (if applicable)

- ✎ Description of penalty
- ✎ Description of conditions under which penalty would be imposed
- ✎ Description of circumstances under which penalty would not be imposed despite existence of penalty conditions such as delayed completion due to no fault of the contractor

Permits

✎ Party responsible for obtaining all building permits

✎ Party responsible for paying permit-related fees

Price

✎ Total price of contract

Royalties and Patents

✎ Who pays for royalties and patent license fees and defense against claims for infringement (normally the contractor)

Safety

✎ Requirements that the contractor and third parties maintain and take precautions for safety and prevention of injury

✎ Requirements that contractor abides by mandated safety laws and regulations

Schedule

✎ Start date

✎ Completion date

✎ Acceptable causes for delay

✎ Procedures for requesting time extensions

Scope of work

✎ Descriptions of the labor, materials, and nature of work to be provided

✎ Description of work that is not included in the contract (See exclusions.)

Supervision

- ✎ Name of persons who will supervise the work
- ✎ Description of the amount of supervision that will be provided

Termination of the contract

- ✎ Conditions under which either party may terminate work and the contract

Warranties

- ✎ Description of warranties

Worker sanitation

- ✎ Provisions for worker sanitation

Zoning

- ✎ Contractor compliance with all applicable regulations

Appendix B

Explanation of Terms

Allowance - A specific and limited sum of money included in a bid or estimate that is to be used for purchasing necessary work or materials, such as cabinet or hardware allowance.

Arbitrator/Arbitrate - A person who is selected to settle a dispute with the consent of both sides. An alternative process of settling a dispute between parties.

Architect - A person whose profession is designing buildings and structures and who has received formal schooling and certification.

Bid - An offer to perform certain work for a specified sum.

Bidding - The process or act of participating in offering to perform certain work or provide certain materials at a stipulated price. Usually in competition with other bidders.

Budget - 1. The total amount of funds available for a construction project. 2. The total estimated or calculated bid amount for a proposed construction project.

Builder - One who builds structures. Also known as a general contractor or contractor.

Building designer - A person who designs buildings and structures but who has not completed either the formal schooling of an architect or certification or both.

Change order - A request either written or verbal to modify or alter the original specifications. A change order can be used to add or deduct certain material or work and may or may not also include changes to the contract price.

Competitive bid - Price quotes for specific work that have been submitted by various contractors with the intent of being awarded the contract based on being the lowest bidder.

Construction drawings - The drawings which illustrate the work to be performed.

Construction team - All parties who have contributed to or will contribute to the construction process and/or completion of the construction; including design, planning, supervising, building, providing materials, etc.

Contract - A document containing the terms of an agreement defining the obligations and responsibilities of each party.

Contract checklist - A list that identifies as a reminder the primary components that might normally appear in a construction contract.

Contractor - One who contracts to supply certain materials or do certain work for a stipulated sum. Also referred to as a builder, general contractor.

Cost plus a fee - A contractual pricing arrangement whereby a contractor provides to the client materials and performs work at the contractor's actual cost plus a predetermined markup of fixed rate as a fee for service.

Cross section - A drawing depicting the interior perspective of a building if it were sliced vertically much like a cake.

Division - A term used to define a broader or least detailed grouping of smaller price components in an estimate. Usually considered to be a larger cost category comprising many smaller and more detailed cost components. (Example: Division of Framing.)

Drywall - Paper covered gypsum board installed over the interior of a building's framing.

Elevation - A drawing of an exterior perspective of the building from one or more sides.

Estimate - A calculation of the approximate cost to complete a general scope of work. Sometimes used interchangeably with bid.

Estimate or bid checklist - A list of cost categories in which all the costs for a proposed project are sorted for the purpose of being able to compare bids side by side.

Financial history - A contractor's payment history to vendors, subcontractors, employees, as well as their financial resources available.

Finish carpentry - The detailed or concluding carpentry work consisting of the interior finish work such as moldings, cabinets, doors, shelving, etc.

Fixed bid - An offer to perform specific work for a stipulated sum.

Floor plan - A drawing showing the footprint of the building along with locations and dimensions of existing and proposed walls.

Framing - The process and or the structural components used to build the structural "frame" of a building.

Interior decorator - A personal with experience or training in the coordination and selection of the interior components and finishes of a building. Also referred to as an interior designer.

Item - The lowest or finest level of detail in an estimate, usually representing a single or detailed cost about individual items or components. Example: item wall studs.

Journeyman - An experienced and fully competent craftsman in a trade.

Laborer - One who performs manual labor which requires no specialized training or skill. Often considered the base or lowest position of service on a job site.

Lead time - The amount of time from the date products or services are ordered to the time they are available for receipt.

Long list - The unscreened preliminary list of potential contractors.

Low bid - The lowest of several cost estimates submitted by a contractor to complete the work defined in the drawings and specifications. Sometimes used as a slang to describe poor quality work, e.g., "The cabinets were obviously low bid."

Lowball - The act with intent to win a contract based on price of knowingly submitting an estimate or bid for an amount which is less than the probable cost to perform certain work or provide certain materials.

Lump sum - A term used to denote a single cost or total cost for a given scope of work or as an allowance for a scope of work.

Minimum requirements - The minimum criteria by which a contractor would still be considered as a candidate for a job.

Negotiated contract - An agreement by which the owner chooses a contractor and negotiates the compensation to be paid to the contractor for services performed.

Not to exceed - A maximum amount which charges for work or materials may not exceed.

Open-ended - The condition of not having any limit or maximum dollar amount or time limit as it relates to construction work.

Paint grade - Material of a quality which is intended to be painted.

Penalty clause - A term used to denote a fine or fee that will be levied against a person or persons in the event certain standards of performance are not met. (Example: schedule.)

Phase - A term used to define a more limited scope or midlevel of price components in an estimate. Typically there will be multiple phases within a division. Examples: Floor framing, wall framing, and roof framing might be phases within a division of framing.

Plot plan - A drawing from a bird's-eye view showing the property, and any existing and proposed buildings, including zoning setbacks, etc.

Preliminary estimate - An approximate cost to perform a preliminary scope of work.

Prequalified - A contractor who meets the preliminary qualifications outlined in the preliminary checklist.

Prescreening - The process of maintaining or eliminating the names of potential builders as candidates for your project based on information gathered through a background check of the builders.

Profit - The difference between the income received from a job and the expenses incurred to complete the job.

Project schedule - The proposed total time anticipated to complete a construction project, often illustrated by a chart or graph indicating the start and completion of key components of the project.

Qualified - A contractor with the proper licensing, insurance, experience, time availability, financial resources, and customer satisfaction history, and who has proven experience with the type of work required for your project, and with whom you feel comfortable working.

Right contractor - A contractor whose skill, experience, personality, availability, and resources fulfill the requirements of your job and your needs as the client.

Roof plan - A drawing of a bird's-eye view of the roof structure and covering.

Rough carpentry - The process of framing or building the structure of a building.

Scope of work - The total of all work to be or proposed to be performed. Usually denoted by the plans and specifications.

Short list - The list of contractors who still remain as possible candidates after screening through preliminary, telephone, and personal interviews.

Specifications - The written instruction or technical drawings describing the exact work or materials that are to be provided or performed.

Stain grade - Material of a higher quality which is intended to be stained and finished.

Subcontractor - One who contracts to supply certain materials or do certain portions of work as part of a larger or primary scope of work.

Supervisor/superintendent - A person in charge of managing or directing the day-to-day activities of a project or scope of work.

Supplier - An individual or business that provides materials or other construction-related supplies and equipment, and who performs no labor on the job. Also known as vendor.

Time and materials - A term used to denote work that is provided or performed on an hourly basis at a specified rate and the providing of materials on an as-needed basis with an additional markup being added for profit.

Unit cost - The cost of a portion of work or material based on a specific quantity of measure. Such as square yard, square foot, or linear foot.

Unqualified - A contractor who for any reason, not necessarily incompetence, does not meet the requirements outlined in the hiring steps.

Value - The fair and satisfactory combination of quality materials, workmanship, price, and service.

Workers compensation insurance - Insurance coverage secured and paid for by the contractor on behalf of his or her employees for their protection in the event of harm or injury while on the job.

ACKNOWLEDGMENTS

Having worked on many and varied projects, I'm again reminded of the benefits of working with a team of talented individuals. Although a good portion of writing a book, even a modest one such as this, requires countless hours of work by oneself, without the expertise and help of many people, the completion of this project would not have been realized.

I am especially grateful to my wife Susan for her patience and support in allowing me to invest the time and resources necessary to complete this project.

A special thanks to Eddy Hall for his advice and skillful editorial expertise. His patience and professionalism made the task of refining the manuscript enjoyable and rewarding.

I would also like to extend my sincere gratitude to each of the following people for their encouragement, expertise, and unbeknown to them, their keeping me accountable for completing this project.

Roger Bolgard	Roger Bolgard, Esq. Legal input and comment Monterey, CA
Charity Bucher	Good Shepherd Publications Page design, composition and print brokering Hillsboro, KS

Joanne Fujii	Copy review
Eddy Hall	Comprehensive editorial services Goessel, KS
Kevin Halloran & staff	Mirthworks Company Advertising and marketing cover design, copy, and review. Honolulu, HI
Roger Mansell	Copy review
Norman Sakamoto	Content comment SM Sakamoto Inc. Hawaii
Allan Freeland	Publishing & Construction Law Cooper-White-Cooper SanFrancisco, CA